Maps in
Tudor England

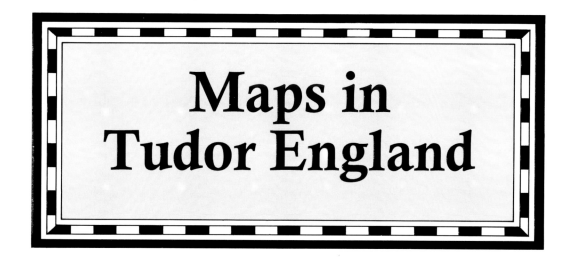

P.D.A. Harvey

THE PUBLIC RECORD OFFICE
AND
THE BRITISH LIBRARY

ACKNOWLEDGEMENTS

I owe a special debt of gratitude to Mr Peter Barber for making available to me, before its publication, his important work on maps and government in 16th-century England. I am grateful to him, too, as well as to Mr Tony Campbell, Miss Margaret Condon and Mrs Sarah Tyacke, for reading drafts of my text and for many helpful comments and suggestions. Dr Elizabeth Hallam Smith at the Public Record Office and Ms Kathleen Houghton and Mr David Way at The British Library have nurtured the work from inception to publication with a courtesy and efficiency which I greatly appreciate.

FRONTISPIECE

Queen Elizabeth I, by Marcus Gheeraerts the younger, c.1592

By the 1590s maps had become such familiar objects that the queen could be portrayed standing on one as a symbol of her kingdom. Her feet rest on Oxfordshire – the portrait commemorates her visit to Sir Henry Lee's house at Ditchley in 1592.

National Portrait Gallery

ENDPAPERS

Mount's Bay, Lizard and St Ives Bay, Cornwall, 1539–40

The westernmost end of a map of the coast from Land's End to Exeter, drawn to plan coastal defences at a time of threatened attack. Another portion of the map is no. 32.

British Library, Cotton MS. Augustus I.i.35, 36, 38, 39

© 1993 P.D.A. Harvey

First published 1993
jointly by the British Library,
Great Russell Street, London WC1B 3DG
and the Public Record Office, Chancery Lane
London WC2A 1LR

British Library Cataloguing in Publication Data
A catalogue record is available from the
British Library

ISBN 0 7123 0311 1

Designed by Andrew Shoolbred
Typeset in Monophoto Calisto
by August Filmsetting, Haydock, St Helens
Printed in Italy

Contents

1 A Cartographic Revolution
6

2 Maps and Fortifications
26

3 Maps and Government
42

4 Maps and Towns
66

5 Maps and Landed Estates
78

6 Maps and Buildings
94

7 Maps and the Law
102

Further Reading
117

Index
119

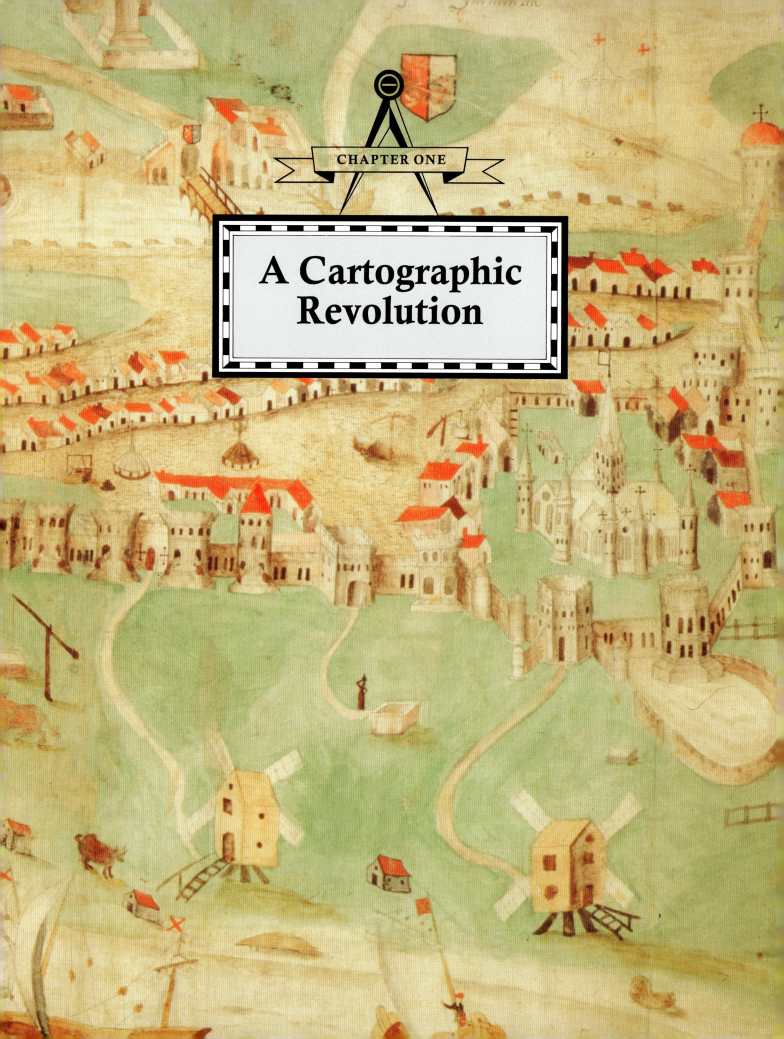

CHAPTER ONE

A Cartographic Revolution

A Cartographic Revolution

IN the England of 1500 maps were little understood or used. By 1600 they were familiar objects of everyday life. When Shakespeare wrote *Henry IV, Part 1* in 1597 he had Mortimer, Glendower and Hotspur divide England and Wales between them by means of a map; these three historical characters in the early fifteenth century could not possibly have used a map in this way, but it is not surprising that Shakespeare – and his audiences – assumed they would. By the 1590s not only were maps consulted for a host of purposes by men of affairs, but they were printed on playing-cards, woven into tapestries, engraved on medals and included as illustrations in Bibles. Queen Elizabeth I was painted standing, symbolically, on a map of England, and some years later, in 1612, the figure of Britannia on the title-page of Michael Drayton's *Poly-Olbion* was shown dressed in a map-like cloak showing woods and trees, hills and rivers.

frontispiece

All this would be unthinkable a hundred years earlier, in the reign of Henry VII. From the twelfth century to the fourteenth there was a peculiarly English genre of world maps and regional maps, probably few in number but certainly unparalleled in contemporary Europe – but this modest tradition had long died out by the end of the fifteenth century. Then the learned may have known the maps of the world and its regions in Ptolemy's *Geography*, which had been printed in Italy and Germany, and they may have known some other world maps. Some English seamen will have seen the portolan charts, used by Mediterranean navigators, which now included the coasts of north-west Europe and west Africa in their outlines. Master masons showed their clients plans of projected buildings. But all these were maps of particular sorts, drawn for particular, limited purposes, and seen by limited groups of people. It hardly ever occurred to anyone to draw even the simplest sketch-map for any other purpose.

The figures speak for themselves. From the second half of the fifteenth century there exist about a dozen maps – or closely linked groups of maps – of particular places or areas in England. From the first half of the sixteenth century we have about 200, from the second half perhaps 800. These last two figures are guesses, but they are informed guesses and are an accurate index of the scale of the change. Perhaps a map drawn in the late sixteenth century would be more likely to survive to the present than one drawn a hundred years earlier – but the difference is probably slight: there are in England remarkably full and well-preserved medieval archives both of private estates and of the government, exactly where these maps would be found if they ever existed. We need not doubt that the pattern of survival is equally the pattern of production and that the sixteenth century saw a cartographic revolution in England.

Nor was it just a revolution in the acceptance and use of maps; it was equally a revolution in the kind of map that was produced. Looking still at maps of small areas – a house, a field, a town, a tract of countryside, an entire county – it is no exaggeration to say that the map as we understand it was effectively an invention of the sixteenth century. Earlier maps might be itinerary-maps, showing the succession of points along one or more routes but not what lay between the routes or how to get from one route to another. Or they might be picture-maps, showing features above ground level pictorially as though seen from a height in bird's-eye view, with more or less attention to detail and varied degrees of artistry – sometimes very little. Or, if they showed only the outlines of buildings or fields, these would be drawn roughly to shape, sometimes with measurements written in, but never to consistent scale. The introduction of scale to these topographic maps in the 1540s was by far the most important of the changes that they underwent in sixteenth-century England. Arguably it was only now with the introduction of isometric drawing, precisely reproducing the results of measured survey, that these depictions of landscape could properly be called maps at all; certainly it was now possible for the first time to distinguish between map and bird's-eye view. By the end of the century maps drawn to scale and showing features by conventional signs were well established. This did not mean, of course, that the older sorts of map were no longer drawn – they still are today – and most of these early scale-maps include many pictorial features. But in the understanding and use of cartography the change was crucial.

So too was the introduction of the printed map. This did not affect the technical development of mapping: the maps printed were of the same kinds as those drawn in manuscript. But it did more than anything else to spread the idea of using maps and plans among the public at large. The first map to be printed in England illustrated the Exodus in a Bible produced at Southwark in 1535, and the maps in many subsequent editions of the English Bible from 1549 onwards will have familiarised people with the idea of maps in general, just as the publication of Christopher Saxton's maps of the counties between 1574 and 1578 will have introduced many people to the idea of consistent scale. Printing ensured that maps were no longer restricted to limited groups of users, to particular techniques and crafts.

It is the maps of small areas – up to the size of a county or comparable area – that are the subject of this book: what they look like, why they were drawn, how they changed in the course of the Tudor period from the accession of Henry VII in 1485 to the death of Elizabeth I in 1603. To exclude maps of the whole country is not an arbitrary distinction. They underwent a parallel development, with successive corrections to the two – somewhat different – outlines that the fifteenth century offered. One of these came from portolan charts, which in the early fourteenth century already presented a fairly correct outline of southern and south-eastern England, based on the actual observation of Mediterranean

A Cartographic Revolution

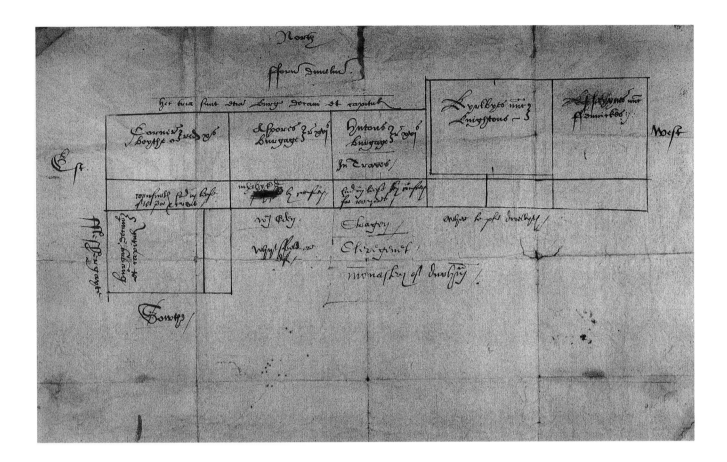

sea-captains who sailed to the English Channel; by the late fifteenth century the portolan charts had achieved a reasonably correct shape for the whole of the British Isles. The other came from the maps, based on geographic coordinates of ancient origin, that accompanied Ptolemy's *Geography*, first translated from Greek into Latin about 1406 and printed in a series of editions from 1475 onwards; overall the outline was rather less accurate than on the portolan charts and, in particular, Scotland protruded far to the east towards Denmark. In the sixteenth century successive maps showed the British Isles in much greater detail and with increasing accuracy: an anonymous manuscript map of 1534–46 and the printed maps of George Lily at Rome (1546) and, especially, Gerard Mercator at Duisburg (1564) mark stages in this development. On what sources they drew is uncertain, but contemporary English local maps may possibly have been among them. However, it was not until Saxton followed his county maps with maps of all England and Wales – to a small scale in 1579, then a large-scale wall map printed from twenty engraved plates in 1583 – that we find an undoubted link between these geographic maps and the topographic

1 House-plots in Durham, mid-sixteenth century

A remarkable plan because, with 'North' at the top and 'Sowthe' at the bottom, it has 'Est' on the left and 'West' on the right. Other sources show that this is no clerical error – the plan really is drawn in mirror image or, put another way, as if the ground was viewed from below instead of from above. This may reflect simple unfamiliarity with the idea of drawing maps.

Muniments of the Dean and Chapter of Durham, Loc.XXXVII, no.113

fig. 2

9

A Cartographic Revolution

2 The British Isles, 1534–46
While in Tudor England the use of maps and plans steadily spread, becoming ever more familiar in every context, geographers were producing increasingly accurate maps of the country as a whole. Here the anonymous mapmaker has rejected Ptolemy's contorted shape for Scotland, but is still far from achieving the correct outline for Wales or Ireland.
British Library, Cotton MS. Augustus I.i.9

3 Newark-on-Trent, Nottinghamshire, and surrounding area, early or mid-sixteenth century
Part of a map that is over 3 feet long, showing the several streams of the River Trent from Nottingham to Newark with the towns, villages and mills alongside. Early Tudor maps are made up of pictures, not necessarily realistic but often, as here, with finely detailed drawing. 'Newarke Town' is bottom right, with 'The castell of newarke' to its left. West is at the top.
British Library, Cotton MS. Augustus I.i.65

A Cartographic Revolution

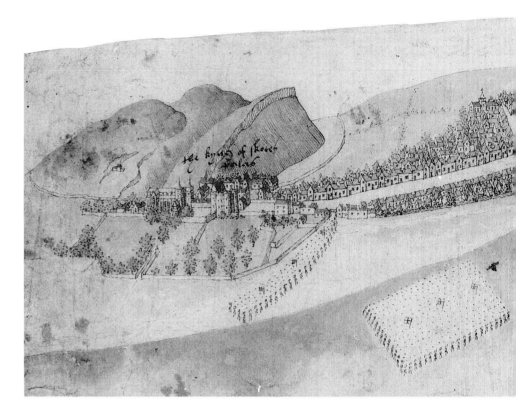

maps of smaller areas. It may well be that until then they developed quite independently of each other.

Nor is it arbitrary to look only at English maps. Inevitably there will be reference to maps of places in Picardy, Wales, Scotland and Ireland, as thrustful English policies meant that English cartography impinged on these areas. But topographic mapping did not develop in exactly the same way in Wales or Scotland or Ireland as it did in England, and in each region of the Continent it followed a separate course – all related, but with important differences in detail. In France, for instance, maps were probably used in law courts long before they were in England; in the Netherlands it seems to have been surveyors, not military engineers, who first drew maps to scale; in southern Germany topographic maps were being printed from the very start of the sixteenth century. We have much still to learn about sixteenth-century mapping in many parts of Europe, and to look at what happened in one cultural area – and England was just that – is a valid contribution to the wider picture. Today we know much more about topographic mapping in Tudor England than we did thirty years ago, and in thirty years' time we shall probably know a great deal more. All that can be done is to give an interim report – a report that is one person's understanding

4 Edinburgh, 1544
Topographic mapping developed differently in England and in Scotland, and the earliest map of Edinburgh was drawn by the English who attacked it under Edward Seymour, earl of Hertford. This part of the map shows some of the English army and the Royal Mile from Edinburgh Castle on the right to Holyrood, 'the kyng of skotes palas', on the left.
British Library, Cotton MS. Augustus I.ii.56

of what happened, one person's understanding of what was significant.

The chapters that follow look in turn at the way maps were used in Tudor England in particular contexts, for particular purposes. At one point or another each function had wider influence, furthering the general acceptance and use of maps, so that overall the book presents a broad chronological account of the way topographic mapping developed in England. Behind this development were, of course, the discovery and adoption of new techniques in drafting and in presentation. Already in the fifteenth century there was increasing use of the compass – or 'dial' – in map-making, reflected in an increasing tendency to draw local maps with north at the top. There was the introduction of triangulation, which was described in Latin by Gemma Frisius in a book published at Louvain in 1533 and in English by William Cuningham in 1559 in his book called *The Cosmographical Glasse*. There was the invention of plane-table and theodolite, whose respective merits surveyors were debating at the end of the sixteenth century. There was increasingly consistent and sophisticated use of uniform conventional symbols – some pictorial, some not – in place of actual pictures on maps, leading to the inclusion of a key on some maps produced in the 1590s.

A Cartographic Revolution

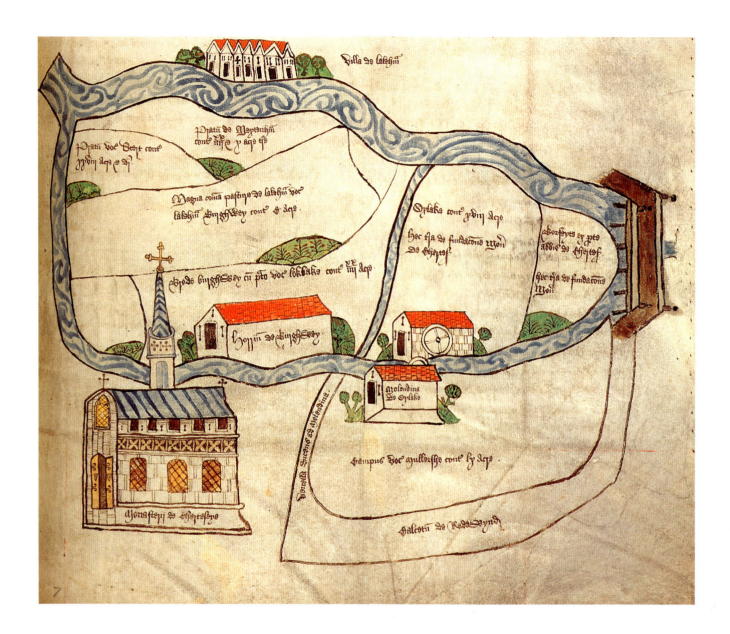

5 Chertsey, Surrey, and Laleham, Middlesex, mid- or late fifteenth century

Drawn in a cartulary of Chertsey Abbey, the map was probably meant as a record of areas where pasture rights had been in dispute. Laleham, on the River Thames, is at the top, and bottom left is Chertsey Abbey on a stream that also flows to Chertsey Bridge on the right, passing two water-mills.

Public Record Office, E164/25, f.222r

6 OPPOSITE   Inclesmoor, Yorkshire, fifteenth century

A later fifteenth-century copy of a map drawn about 1407 for the Duchy of Lancaster, showing an area where rights of cutting peat were disputed with St Mary's Abbey, York. At the north-east corner (bottom left) the Trent, Don and Ouse all flow into the Humber; Ferrybridge, on the Aire, is at the west end of the map (centre right). The details of houses, bridges, wayside crosses and marshland plants make this a unique view of the medieval rural landscape.

Public Record Office, MPC 56

A   C a r t o g r a p h i c   R e v o l u t i o n

But what made the cartographic revolution of the sixteenth century was not simply the discovery and acceptance of new techniques. Rather it was the acceptance and spread of unfamiliar concepts. Far more than a revolution in the ways maps were made it was a revolution in the ways of thought of those who used them. More and more people discovered what a map was and appreciated how useful it could be in ever more varied circumstances, achieving understanding first of the relatively straightforward picture-map or diagram, then of the map drawn to scale – a more sophisticated, artificial construct. Cartographic techniques were substantially in advance of the market in Tudor England, ready to be put to use when demand arose. What mattered was the spread of demand, and how map-makers created and fostered this demand for

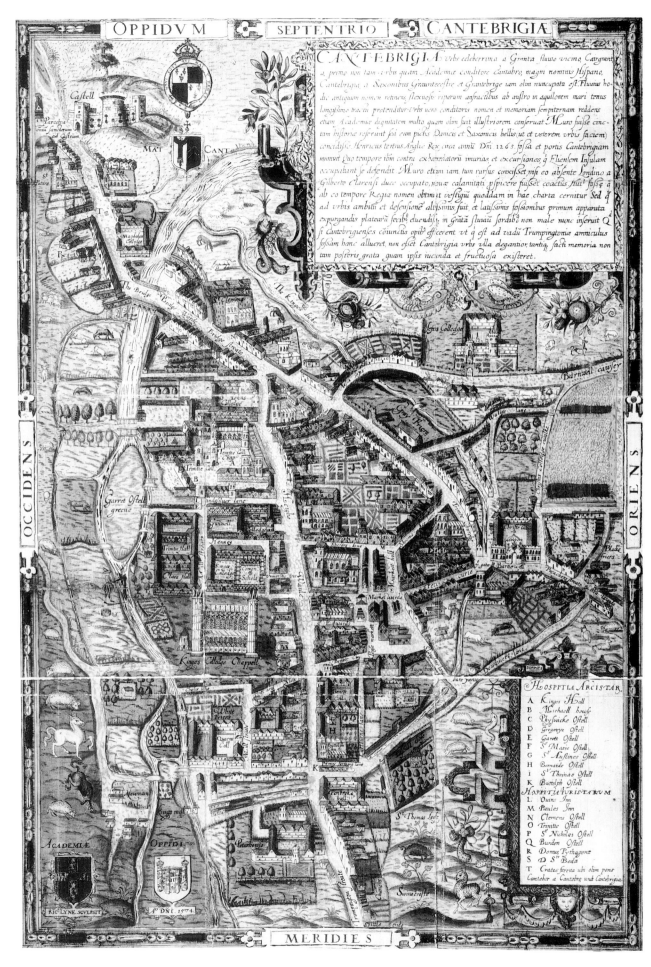

7 Cambridge, 1574

At first sight a bird's-eye view, but in fact a conventional map, to consistent scale, with pictures superimposed on it, a technique often used in the sixteenth century. At the top, beside the castle, are the arms of the queen and of Matthew Parker, the archbishop of Canterbury; bottom left are the arms of the university and the city with the date and the name of the engraver, Richard Lyne. The map illustrates a book arguing that Cambridge University is older than Oxford, and at the top is a long historical note.

British Library, C.24.a.27

8 Fields near Andwell, Hampshire, early sixteenth century

Really a picture from an imaginary vantage-point, this is an extraordinarily evocative everyday scene set in a winter landscape. In the foreground the pack-horses and cart are moving along 'The hie Wey ledyng from london to Basyngstoke'; behind are 'Hurste barne' and fields and woods with notes of their names and areas.

Muniments of Winchester College, MS. 3233

their products. Certain map-makers played a crucial role in this, but no less crucial was the role of their customers. Every map had two parents, the map-maker who drew it and the customer or patron who commissioned it and paid for it, and in looking at the history of maps in Tudor England we shall be as much concerned with the patrons as with the map-makers. It would have been of little consequence that John Rogers and Christopher Saxton and Ralph Treswell were able to draw maps if Henry VIII and Thomas Seckford and Sir Christopher Hatton had not been prepared to pay them to do so. What we see is successive extensions of an awareness of maps – of mapmindedness – until by the end of the sixteenth century the map-maker could see any literate person as a potential customer. It was a revolution in cartographic understanding.

The maps reproduced throughout the book have been chosen to illustrate the development and spread of cartography. But they tell another story too. Nearly all the early Tudor maps are, simply, pictures, and even when, with the introduction of scale, measured ground-plans became the basis of many maps, most maps still retained pictorial features, sometimes stylised but often drawn from life. Indeed, a form of map developed which superimposed on a scale-map pictures of houses, trees and so on, each drawn individually in perspective. This technique was used particularly for town plans, and Richard Lyne's map of Cambridge in 1574 is an unusually clear example; sometimes it is difficult to tell

fig. 7

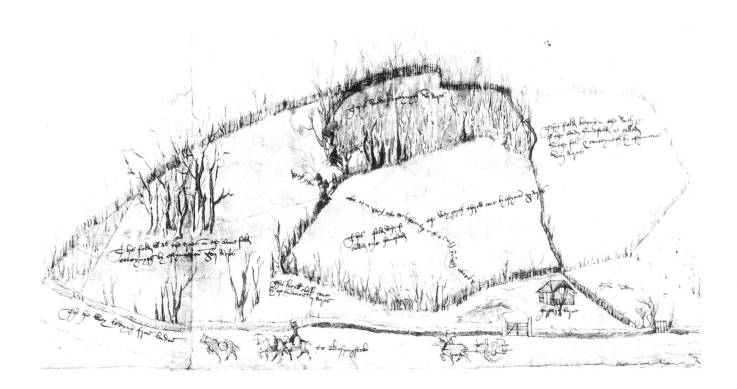

A Cartographic Revolution

A Cartographic Revolution

9 Great Yarmouth, Norfolk, mid- or late sixteenth century
A detailed view of the town with a lively if selective picture of activities going on around it. West is at the top, the sea in the foreground. The prominence of the river mouth and its embankments (left), showing works carried out in 1566, suggests that the map was meant as a guide to planning these or further works. Another possibility is that it was copied from a map drawn to decorate a civic room – the borough arms appear top right.
British Library, Cotton MS. Augustus I.i.74

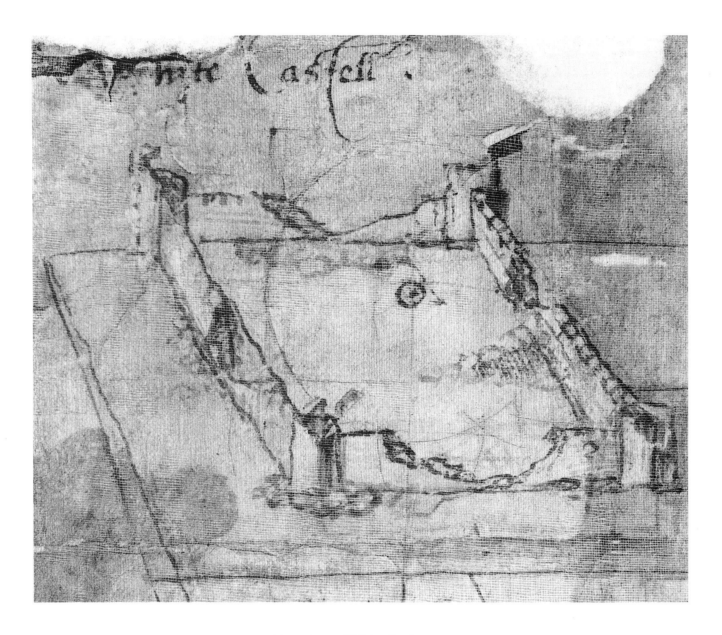

10, 11 Two views of White Castle, Monmouthshire, late sixteenth century

These are contemporary but very different representations of the same structure, each a tiny picture on a map in the archives of the Duchy of Lancaster, the castle's owner. They show how cautious we must be in using as evidence all the pictorial detail we find on Tudor maps – they were interpretative drawings, not photographs.

10 Ruinous curtain walls probably reflect accurately the actual state of the castle in the late sixteenth century, but their rectangular shape is incorrect.
Public Record Office, MPC 36

11 OPPOSITE   Though this must show the castle as it once had been, not as it was at the time, its general shape and appearance are much closer to reality.
Public Record Office, MPC 93

whether features have been imposed on a base drawn to uniform scale or whether the whole is a genuine perspective drawing in near-vertical bird's-eye view. However, one way or another all these maps give us pictures of the landscape of Tudor England.

This is the more important because there was no other English genre of landscape drawing in the sixteenth century. Nor do we have more than a very few portrayals of specific landscapes from the middle ages – indeed, hardly any apart from the handful of medieval picture-maps, which in the fifteenth century include views of Chertsey Abbey and of the rural landscape of Inclesmoor in south Yorkshire. The view of the Tower of London, with London Bridge and the city in the background, painted about 1480 to illustrate a copy of the poems of Charles, duc d'Orléans, was a most unusual production. Effectively it is the pictorial maps of Tudor England that give us our first glimpse of what England actually looked like in the past, and this they do extraordinarily well. For one thing they are extremely varied in style. On one hand we have the quaint and artistically primitive picture of, for instance, the shore of the Thames estuary in about 1514. On the other we have the contemporary but vastly more accomplished drawing of fields, woods and the road from London to Basingstoke near Andwell in Hampshire, one of the most impressive and evocative of all these views of Tudor landscape, with its wintry trees and hedges, fences and a field-barn all drawn with precision and artistic skill. But there is no less variety in subject matter, for as the use of maps spread they brought within their purview every kind of landscape. The Andwell picture-map is of a fairly nondescript piece of countryside. There could hardly be a greater contrast with the map of London printed in the 1550s, just as precisely drawn and scarcely less evocative, with its crowded mass of streets, houses and churches, its riverside warehouses, wharves and cranes, and outside the city wall its suburban houses and gardens, archery butts and grounds for drying cloth. Moreover many of the maps people the landscape, showing the scene as it really was, the landscape in use. Some do not – Ralph Agas's magnificent estate map of Toddington in Bedfordshire in 1581, which includes the finest view we have of an Elizabethan village, shows the scene almost devoid of inhabitants. But in the Andwell map we see packhorses and a horse and cart making their way along the road, and some maps show us a host of people going about their business, whether in the operations of war or in the peaceful affairs of everyday life. In the late-sixteenth-century map of Great Yarmouth and in the London map of the 1550s there are many groups of figures, riding, carrying water, shooting with bows and arrows, and so on.

It is interesting, though, that in each of these maps the figures are all in the roads and grounds outside the town. The streets of the town itself, even the house-lined streets of the suburbs, are empty. This raises an important point. The Tudor picture-maps present a magnificent panorama of contemporary

A Cartographic Revolution

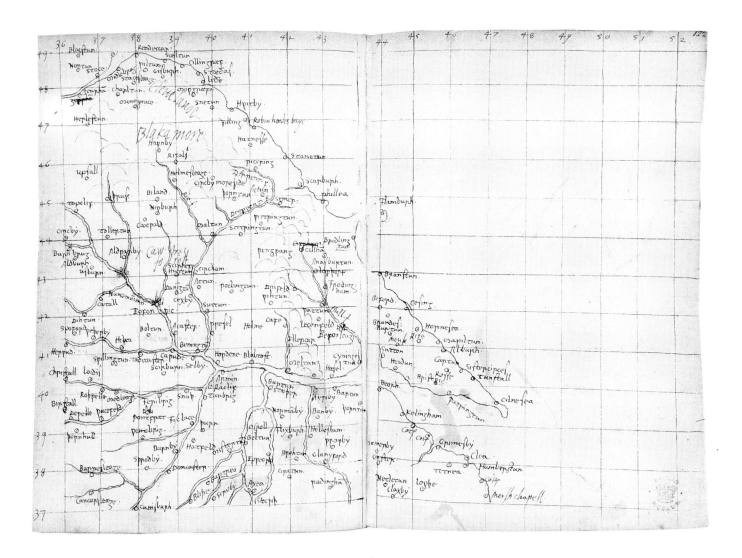

12 The Humber and east Yorkshire, by Lawrence Nowell, mid-sixteenth century

One of thirteen sections of a map covering all England and Wales. Nowell was an Anglo-Saxon scholar as well as a map-maker, and he not only gives place-names in Old English – as 'Eoforwic' for York – but uses Old English letter forms. We see this in the sequence of names along the coast from the north – 'Readeccar', 'Sceltun', 'Cillingraef' (Skinningrove), 'Staethas', 'Lithe', 'Hwitby'. Interest in antiquities and in map-making often went together in the sixteenth century.

British Library, Cotton MS. Domitian A.xviii, ff.121v–122

landscape, but they are not objective pictures like photographs. This does not diminish their value – insofar as it can be properly understood it vastly increases it, for we learn not only what sixteenth-century England looked like, but how it appeared to those who lived in it. Probably the Great Yarmouth and London maps show people around the town but not within it simply because the streets were already full of detail, and innumerable figures would have overcrowded and confused the picture. Outside the walls, however, there were awkward blank spaces; the figures served to fill them and to present a balanced composition, adding interest to what would otherwise be a lifeless representation. But they were not drawn true to life, for – we may reasonably guess – there would really have been far more people to be seen around the town. Those that appear are representatives of a much greater number and have been chosen because

A Cartographic Revolution

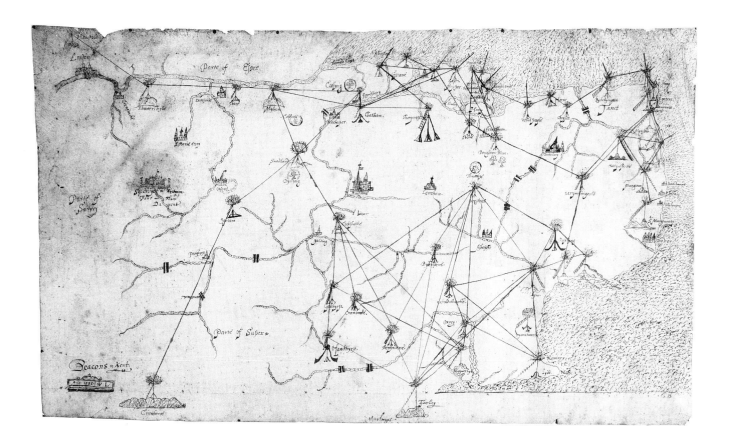

13 The beacon network in Kent, by William Lambarde, 1585

The beacons were used when local forces needed to be raised urgently and were constantly manned when foreign attack threatened. The map shows the complexity of the system around the coast and the lines of visibility between beacons, and we see how Christopher Saxton may have used the beacons as triangulation points when he mapped the whole country in the 1570s. Lambarde, legal writer, historian and archivist, was (like Lawrence Nowell) both antiquary and map-maker.
British Library, Additional MS. 62935

they were doing things that seemed especially characteristic or especially interesting. Even if they were drawn only to fill a blank space, their selection tells us something of the way the map-maker saw the scene. Viewed in this light, it does not matter that some of the picture-maps are naively drawn, or that some of them show features in conventionalised form: they reflect just as well – or even better – the map-maker's perception of landscape, his reaction to it. Occasionally we see how this perception differed from one map-maker to another. Two late-sixteenth-century maps in the Duchy of Lancaster's records show parts of Monmouthshire and include a tiny picture of White Castle. One shows the towers and battlements of a well fortified stronghold; the other shows a ruinous curtain-wall with no buildings inside. No doubt one was conventionalised ('this is how a castle ought to be'), the other realistic ('this is how it actually is') – but both are valid reactions of the map-maker to a prominent feature of the local landscape.

figs 10, 11

These differences in perception are more easily understood if we appreciate that maps were drawn for many different reasons in Tudor England, and that a single map might serve more than its obvious primary purpose. It is proper as

23

well as convenient to examine their development under the heads of various practical functions – fortification, estate administration and so on. But this can obscure much of the point of their spread and acceptance. Christopher Saxton's county maps of 1574–8 were of obvious practical use, for national and local administration, for travellers, for general reference – they are a cartographic guide to England and Wales. But there was more to them than that. On each map are decorations: there is an ornamental border, there are the arms of the queen and of Thomas Seckford who commissioned the maps, there are cartouches with elaborate strapwork and architectural detail, fruits and flowers. We should not see these as extraneous, irrelevant appendages; they are an integral part of the map, for it was meant to be a decorative artefact as well as a practical one. When we hang one on our wall because it is ornamental we are doing exactly what the map-maker intended. But the county map had still more to say. The map symbolised the county, was a statement of its identity, just as the map that Elizabeth I was standing on in her portrait symbolised the country as a whole. This too is a function that it may still serve today, when the early maps people hang on their walls are maps of their own native county – it is a statement of their origin and roots. But in the sixteenth century attachment to the county and its society was stronger, and the map as an expression of it was a more potent symbol. We do not have to explore deeply the psychology of art or of cartography to see that a map, like a picture, may have more than one level of meaning, more than one message – conscious or subconscious – for those for whom it was made.

frontispiece

This kind of symbolic statement may have been the chief *raison d'être* of some Tudor maps. We cannot suppose that the map of London in the 1550s was printed to help strangers find their way round the city. Rather, like contemporary maps and views of important cities on the Continent, it was for display as a statement of London's size and wealth, an expression of civic pride. Still more, pride of possession, the display of a statement of ownership, was probably what impelled many estate owners to commission maps of their properties, quite as much as the simple utility of the finished product. Even where the functional value of a map was clearly paramount its message may be complex. A map drawn by a military engineer or drawn to present to a court of law may show simply a particular fortification or a particular tract of country, but it will have been drawn in such a way as to illustrate or emphasise the advantages of a scheme of defence or the merits of a claim to property. In looking at any early map we must bear in mind all these connotations, all these hidden messages.

Nor was Tudor England's idea of practical value necessarily the same as ours. Thus it would be quite wrong to think of the maps in Bibles as lying outside the sphere of practical activity: they are an expression of Protestant attitudes to the Scriptures, and all religion, whether Protestant or Catholic, lay close to the heart of contemporary politics and diplomacy. One theme that constantly recurs in Renaissance cartography is the involvement of antiquaries and historians in map-making. Early Italian city plans give particular prominence to ancient monuments and the first printed map of the kingdom of Naples, in 1556 or 1557, includes the classical Latin forms of place-names and names the ancient Italic tribes. English antiquaries' connections with mapping began long before the rule of the Tudors and continued long after – in the early fifteenth century Thomas Elmham included a map of Thanet in his history of St Augustine's Abbey, Canterbury, and William Dugdale himself drew the maps to illustrate his *Antiquities of Warwickshire*, published in 1656. In the sixteenth century, Lawrence Nowell used Old English lettering on his map of England, and William Smith, herald and antiquary, was a notable map-maker. To all this too there was a practical side. The possession of landed property – and disputes about it – would often depend on ancient records and their interpretation. Civic rights might rest on early grants and privileges. The antiquary might thus play a valued role in affairs and had close links with both lawyer and surveyor. Religion, ancient records, heraldry, were as much practical matters as fortification and architecture – they were all part of the structure of society. It was this society which, in the course of the sixteenth century, discovered the value of maps.

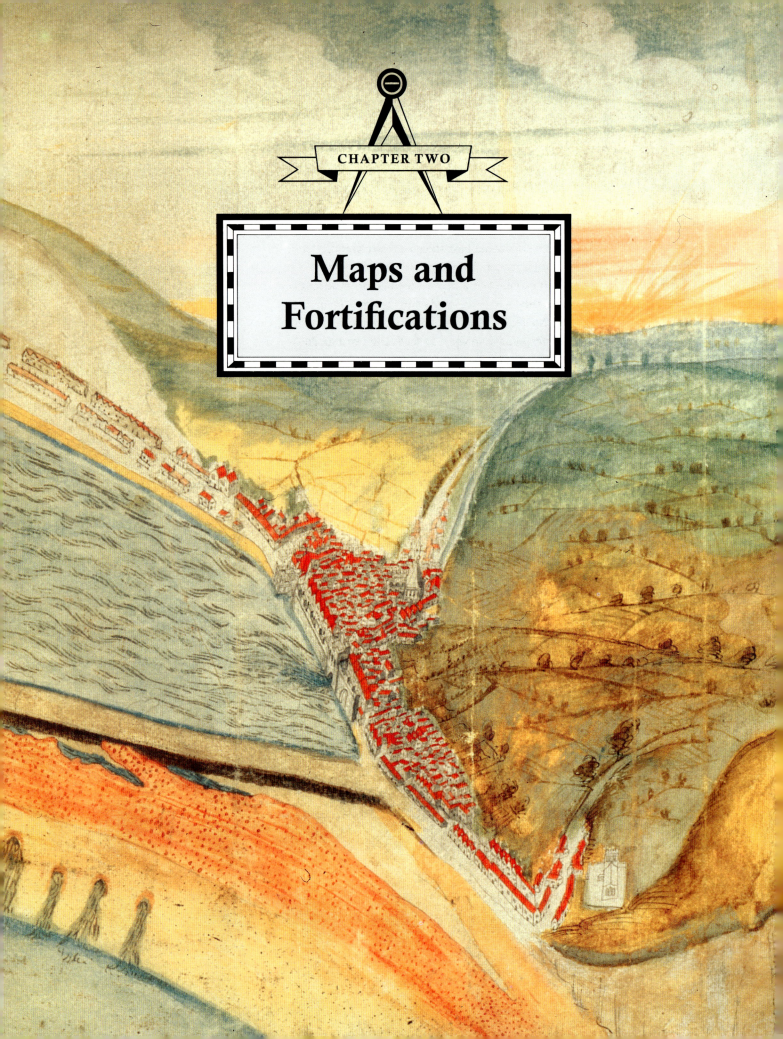

CHAPTER TWO

Maps and Fortifications

Maps and Fortifications

THE cartographic revolution in England began probably at the very start of the sixteenth century. It was not until the 1530s, however, that there was radical change in any one field of map-making. Rather there seems to have been a sudden spread of the idea of drawing maps – picture-maps and diagrams – for various purposes and in various contexts. In 1508 Sir John Ferrers and 26 others sealed – some also signed – a declaration that a meadow claimed both by the bishop of Coventry and Lichfield and by the manorial lord of Elford in Staffordshire was part of Elford manor. 'And for the more playne declaracion of trouthe we ... haf made and causyd to be made a platte or a tervete of the said Maner of Elford with thappurtenaunces and the bondes (*i.e. bounds*) of the same' – this map was attached to the written document. *Tervete* is an unusual word for what must in 1508 have still been an unusual object; but *plat* was to become a normal word in sixteenth-century England for perspective views, picture-maps, scale-maps, ground-plans – topographic drawings of every kind. There is no parallel a generation earlier to drawing a map in such circumstances.

fig. 14

But although this instance reflects and illustrates the way the use of maps suddenly spread in the early sixteenth century, it is unusual in being precisely dated. More typical is a picture-map of three mills near Wimborne St Giles in Dorset. The mills, lands and streams are named, two buildings mark the villages of Gussage All Saints and 'Gussage Bounde' (Gussage St Michael), and it may well have been drawn in connection with a dispute over the flow of water to the mill farthest downstream. There is probably enough information on the map for it to be placed in its context and assigned a more or less precise date; but this could well involve lengthy and complex research on surviving records of the area. All we can say – as of many other maps – is that it looks as if it was drawn in the early sixteenth century. Even if we had a complete census of the topographic maps from Tudor England we should still be in the dark over much of the chronology of their development; this could be established only by patient, detailed research into the local historical context of innumerable individual maps. We can discuss with confidence where and how maps were used in planning fortifications only because so much research has been done on these and other royal works by H.M. Colvin and his associates in the *History of the King's Works*; they have supplied a firm chronological basis for the accompanying cartographic development.

fig. 16

The systematic use of maps in planning these royal works seems to have been the first radical extension of cartography in Tudor England – it introduced maps into an activity that had hitherto managed without them. The new styles of fortifying towns and citadels, which were to dominate military thinking for the next 200 years, were based on systems of angle bastions that in an age of gunpowder offered the strongest defence. They developed in Italy at the end of the fifteenth century, but were soon adopted throughout Europe, and the earliest published treatise – the first of many – was by Albrecht Dürer and appeared in 1527 at Nuremberg. Dürer illustrated his work with plans, elevations and profiles, and graphic representation, whether in plans or in views – and in England all would have been 'plats' – seems to have gone hand in hand with fortification throughout early-sixteenth-century Europe. One of its most impressive relics is the *Livro das Fortalezas* which Duarte de Armas produced between about 1509 and 1516: two views and a plan of each of 57 fortresses along the Portuguese frontier. It is the more remarkable in that in Spain and Portugal there was probably even less precedent for drawing topographic maps than in contemporary England – the idea may well have come from Italy.

In England we get a first hint of the association of plans with royal works in two commissions to Vincenzo Volpe, king's painter. The first was from the king himself, in 1530, to make a plat of Rye and Hastings – the defences of Rye were being strengthened at the time. The second was from the Corporation of Dover in 1532, to show the king a plan for a new harbour as the old one was blocked by silting. The first is lost, but 'A Plott for the making of the Haven of Dover' survives, a very fine bird's-eye view. Volpe was indeed an artist, one of a series of Italian sculptors and painters employed at the English court throughout Henry VIII's reign. Another Italian artist who entered the king's service, in 1538, was Girolamo da Treviso, employed, according to Vasari's *Lives of the Painters*, 'not so much as a painter, but rather as an engineer'. He was indeed a man of parts. At Bologna he had painted a Madonna and saints for the church of San Domenico; for Henry VIII he painted an anti-papal allegory, showing the four evangelists stoning the pope. He built himself a house at the king's expense; and in the king's service he produced a report, with map, on the defences of Montreuil in Picardy, which was criticised as seriously misleading. He defended himself, but the king's advisers felt that he was inexperienced in sieges and unskilled in the details of fortification. Finally, at his own request, he was put in command of a troop of a hundred arquebusiers – musketeers – and was killed in the assault on Boulogne in September 1544.

fig. 17

Da Treviso was unlucky, and perhaps incompetent as well, but the combination of artist and engineer was far from unique – Dürer, after all, was another, much better, artist who was interested in military works, and so too was Leonardo da Vinci. After Da Treviso, several other Italian military engineers were employed by Henry VIII in the 1540s and we might see Da Treviso as bridging a gap between Volpe, who drew views as an artist, and the engineers whose plats would be more technical. In fact, though they mostly lack Volpe's artistry, surviving maps that can be linked with these Italian engineers were nearly all picture-maps as before, often quaint and following rules neither of perspective nor of scale. In 1545 Antonio Bergamo signed alterations made to a plan of

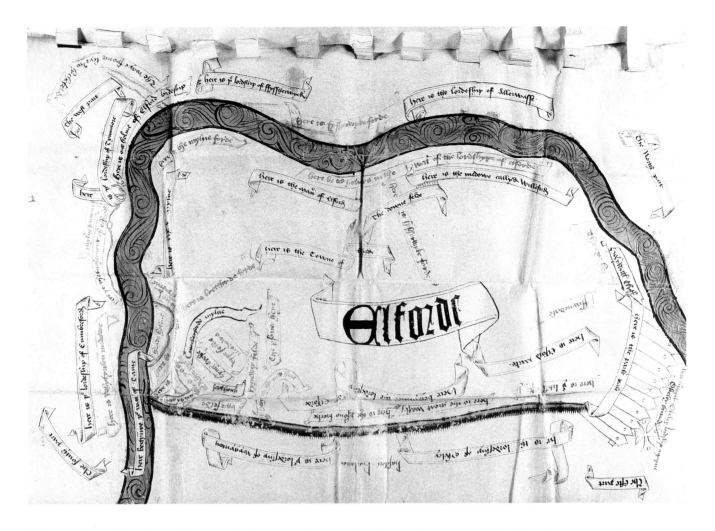

fig. 47

Wark-on-Tweed Castle, and Archangelo Arcana, whom the Scottish ambassador described as 'the best gunner in England', drew a plat of Kelso that has not survived, while maps of Newcastle-upon-Tyne and Tynemouth of about the same time with inscriptions in Italian are believed to have been by Gian Tomasso Scala. One surviving anonymous map of Boulogne and neighbouring forts that was drawn for the English king is also inscribed in Italian. Henry VIII was not alone in employing Italian engineers at this time; others were in the service of the king of France. But Henry also drew his engineers from other parts of Europe: one, from Portugal, is thought to have drawn a map of Guines with inscriptions in Spanish and French, while Stefan von Haschenperg, a Moravian, was in charge of royal work at Carlisle from 1541 to 1543 and drew three surviving plats of its castle and walls. Of all surviving plans known to have been drawn by these foreign experts only two, by Giovanni di Rossetti, showing fortresses at Ardres in Picardy and at Broughton Craig in Angus, seem to be drawn to scale.

The plan of Broughton Craig was drawn in 1547. By then English engineers were fully familiar with the idea of drawing plans to consistent scale. From their chronology it is possible that Da Treviso introduced the technique, but

14 Elford, Staffordshire, 1508
A diagrammatic plan that is attached to a formal sealed statement in a dispute over the ownership of 'the medowe callyd Willeford', shown upper right beside the River Tame. Only the neatly written inscriptions on scrolls with hatching were originally on the map – the many other scrolls with wording were added later. North is at the top right corner.
Birmingham Reference Library, Elford Hall MS. 55

even if it came from Italy there were other possible channels. A treatise by Niccolo Tartaglia, first published at Venice in 1538, describes various technical subjects in a series of imaginary conversations, and in the chapter that tells how to draw plans to scale the dialogue is with 'Mr Richard Wentworth, gentleman of his majesty the king of England'. This Richard Wentworth has not been identified and may be fictitious, but at least the treatise shows that to Italians in the 1530s the idea of English courtiers coming to learn the latest techniques was not a strange one. However, the concept of the scale-plan may have reached England from some other source. Some of the plans in Dürer's treatise of 1527 are drawn to scale – 'measured by the little foot', that is, corresponding to the full-size foot on the ground – and in the Netherlands surveyors were using both the phrase and the technique by the 1530s.

The earliest maps drawn to scale by English map-makers date from about 1540. They are not, however, of anywhere in England, but of Guines, in the territories around Calais that the English had held for 200 years. The castle and other defences were being strengthened by Richard Lee with the assistance of John Rogers, both of them master masons employed by the king as military engineers. Of the surviving plats made for these works, four are plans drawn to consistent scale, two explicitly so: one has a small scale-bar, the other has a note 'The Inshe conteynyth L fotte' with a continuous scale-bar, marked in 20-foot divisions, that almost frames the entire map, evidence of the need to explain the idea of the scale-drawing to the map user who would be unfamiliar with it. Three of these four plans were drawn by Rogers himself or, possibly, by Lee. Another scale-plan, drawn in 1541, shows an area near Calais, recently drained, where a new settlement was planned, and it is especially interesting because of its precise notes of direction beside several roads – 'est & ii pontes (*i.e. points*) by southe' and so on. Drawing maps to scale involved the exact measurement not only of lengths but of angles as well.

fig. 20

fig. 21

Our earliest scale-plans from England itself were drawn by Rogers soon after October 1541: three plans of fortifications near Hull. Two have scale-bars, the other has notes of measurements. Although the central feature is in plan, all three show some features pictorially. Thus on one we have a detailed ground-plan of defensive towers and the wall between them, while across the River Hull the town's houses lining the water are drawn in perspective with smoke coming from their chimneys. Another, covering a wider area, shows roads and streams west of Hull carefully drawn to scale, but marks the villages by tiny pictures of houses and at the top (the west) the map has a skyline of receding ranges of hills. Other maps of the early 1540s show fortifications at Carlisle, Dover and Harwich drawn to scale, and other scale-maps come from the territories around Calais between 1545 and 1547, showing how defences at Ambleteuse and Boulogne were to be improved. In all some 40 scale-maps of English origin

fig. 25

15 Dover, Kent, 1581
The harbour before the new works of the 1580s. These were under the command of Thomas Digges, whose arms appear on the left. At least the harbour and its surrounds are drawn from measured survey, and the pictures of buildings superimposed. The notes of soundings and anchorages have been added later. The harbour works at Dover were, like fortifications, undertaken by the Crown's engineers.
British Library, Additional MS. 11815a

survive from the 1540s; about half are of places in Picardy, half of places in England. Some have scale-bars, but often the scale is given in a phrase which explains the idea of scale – thus on a map of Thanet in 1548 we read that 'In thys plat 3 inches contayneth one mile'. As more of these scale-maps were drawn this was no longer necessary, and a Latin note above the scale-bar in a 1572 map of Sheppey reads simply 'Scala passus bis Mille in Semipede' – scale 2000 yards to half a foot. But, as we shall see, this did not mean that maps drawn to scale were generally accepted and understood. The plans of fortifications were made for the king, his ministers and military advisers – a strictly limited circle of knowledgeable people, who might be expected to understand the techniques of engineers' plans as well as the techniques of fortification itself.

fig. 22

In any case the plats of fortifications made in the second half of the sixteenth century were not necessarily scale-plans of the sort drawn by Lee and Rogers. At Berwick-upon-Tweed, continuing work on the fortifications in the 1560s produced not only ground-plans drawn carefully to scale, but also, about 1570, a magnificent bird's-eye view of the town and its walls. At the other end of England, we have from Dover at least seventeen plats stemming from successive schemes for the harbour and its defences in the sixteenth century. The earliest are two splendid views, the one drawn by Vincenzo Volpe in 1532 and the other, possibly by Richard Lee, submitted to the king in 1538 by John

fig. 17

Maps and Fortifications

16 Wimborne St Giles, Dorset, early sixteenth century
The date and purpose of this map are unknown, though it was probably drawn to show the sources of water for the three mills. The two towers on the lower stream represent the neighbouring villages of Gussage St Michael and Gussage All Saints.
Nottingham University Library, MiP 14

fig. 18

fig. 15

fig. 19

Thompson, surveyor of the works there. Later projects produced plans drawn to scale in whole or in part, such as one drawn in 1581 and embellished with the arms of Thomas Digges, the engineer in charge – here the buildings of the town and castle are drawn in perspective but superimposed on a scale-map. But they also produced a plat of 'new worke to be done and already done' at some point later in the century which has been called 'garish' with its near-Impressionist view of the sun setting over the landscape behind the town.

Detailed work on the styles and hands of the many anonymous plats of fortifications in the sixteenth century might identify more individual mapmakers, showing what different sorts of map each draughtsman made and how they developed in the course of time. Probably, though, this would be difficult and the conclusions only tentative. In analysing John Rogers's short career as a map-maker, from 1540 to his death in 1558, L.R. Shelby found some inconsistency in style: without the evidence of handwriting some of his plats would not have been attributed to the same draughtsman. On the other hand he found that Rogers's work improved over the years, not in its architectural drawing – as a master mason he would already have been practised in this – but in the ways it represented landscape, something he was probably attempting for the first time.

It was not only in drawing their maps to scale that Lee and Rogers were the forerunners of a new trend in English fortifications and other royal works. After

33

Maps and Fortifications

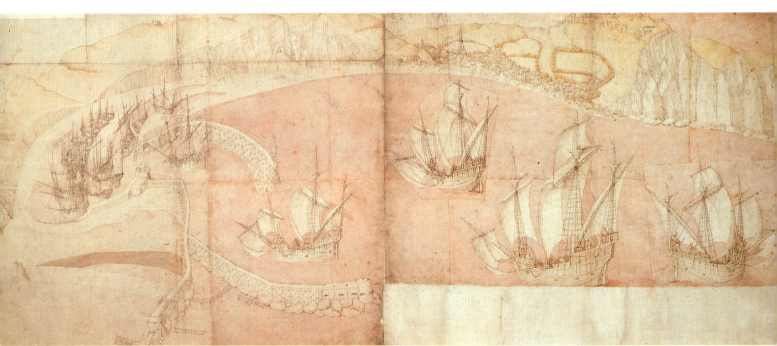

17–19 Dover, Kent, 1532 – c.1590
Three plats, in contrasting pictorial styles, that were drawn for schemes for works in Dover Harbour. They can be compared with the scale-plan of 1581 (no.15). Dover Harbour was constantly at risk from storm and from blockage by sand and pebbles deposited by the coastal current. From 1532 the Crown undertook a series of projects to meet these problems, but it was not until 1585 that a fully successful scheme was completed.

17 OPPOSITE, ABOVE Drawn by Vincenzo Volpe, the king's painter, in 1532. It was commissioned by Dover Corporation to present to the king, and shows an ambitious scheme for a new harbour behind the shoreline.
British Library, Cotton MS. Augustus I.i.19

18 OPPOSITE, BELOW Presented to the king in 1538 by John Thompson, in charge of the works at Dover. It shows on the left the proposed – ultimately unsuccessful – scheme, but does not distinguish between what was complete and what remained to be done.
British Library, Cotton MS. Augustus I.i.22,23

19 The origin and purpose of this extraordinary picture are not known, but from the form of the harbour it probably dates from about 1590. Although by now scale-plans were normally produced for engineering works, old-style pictorial plats might still be drawn.
British Library, Cotton MS. Augustus I.i.45

Maps and Fortifications

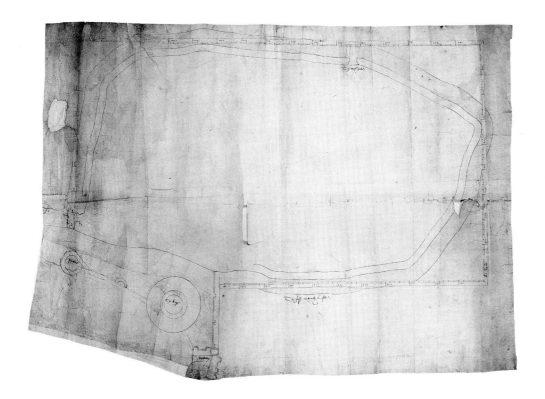

20 Guines, Picardy, *c*.1540
A plan of the existing walls and towers at Guines, preparatory to planning new defensive works. This is the earliest topographic map drawn to scale by an English map-maker. It is the work of either Richard Lee or John Rogers, who were in charge of fortifications in this outpost of the English territories around Calais. The scale-bar almost surrounds the map and there is also a note that 'The Inshe conteynyth L fotte' – early scale-maps often have some such explanation of scale for those unfamiliar with the idea.
British Library, Cotton MS. Augustus I, Suppl. 14

21 Projected settlement near Calais, 1541
A newly drained area divided into plots of from 30 to 200 acres, with a house on each and a church at the centre – comparison with the later scheme for a lordship in Ireland (no.41) is interesting. Notes of compass directions along the principal roads show that it was surveyed and measured on the ground and – though there is no scale-bar – drawn to scale.
British Library, Cotton MS. Augustus I.ii.69

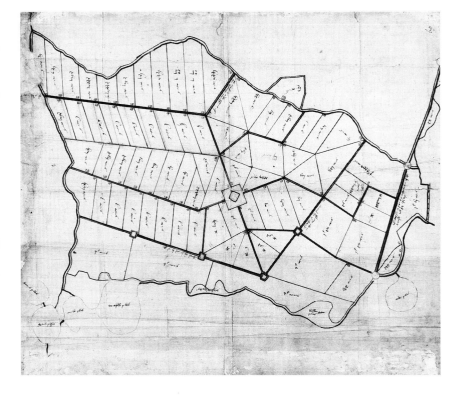

the 1540s the engineers in the king's service – and thus the map-makers – were English, with very few exceptions: Federico Genebelli, who wrote inscriptions on his maps in Italian, Fernando Poyntz, who came to England from the Low Countries, and one or two others. The Englishmen brought diverse skills to their work. Richard Popinjay, who was in charge of works at Portsmouth and elsewhere on the south coast for most of Elizabeth's reign, was a military engineer by profession and nothing else, but Arthur Gregory was an inventor and cryptographer, skilled in the secret unsealing of letters, and Thomas Digges, in charge of the brilliantly successful scheme for Dover Harbour in the 1580s, was a mathematician whose father, Leonard, had written a treatise on surveying. His predecessor at Dover, John Thompson, whose design for the harbour 50 years earlier had been decidedly less successful, had been a local clergyman. Although many were experienced in building of every kind we no longer find the earlier combination of artist and engineer – and plans of new works were increasingly seen as matters of accurate measurement and drawing, albeit with some pictures added, rather than as exercises in unfettered artistic skill.

The production of plats of fortifications and defensive schemes was always spasmodic. It reflected fluctuating royal finances as well as varying concern over defences, greater in times of war and national emergency than during peace, and at any one time work was concentrated on places considered especially vulnerable. We have already seen how particular works in Picardy give rise to maps of Guines in about 1540 and of Ambleteuse and Boulogne in 1545–7. Surviving maps from Plymouth and Portsmouth illustrate this pattern and the sort of contrast found between one place and another. Broadly, ignoring the hesitations and niceties of approximate dating, we have from Plymouth one map from the late 1540s, one from the mid-1550s, one from the late 1580s, eleven from 1590–1602 and several more from the first half of the seventeenth century. From Portsmouth we have one of about 1539, one from 1545–6, one from 1552, six – all by Popinjay – from 1563–77, twelve from 1584–7, then none at all until 1662. We shall see in the next chapter how such an emergency as the expected attack of the Spanish Armada in 1588 produced not only plans of individual works, hurried forward, but other maps of broader defensive strategy as well.

Some of the plats of fortifications cover a wider area, looking at the particular work in its local strategic setting. Two early examples already mentioned are John Rogers's map of the country behind Hull in 1541 and the map of Thanet in 1548. At Dover not only the picture-maps in older style, but also the scale-map drawn for Thomas Digges in 1581 show the town, the castle and their surrounds as well as the harbour works that are the actual subject of the map. One of Rowland Johnson's maps for Berwick in the 1560s extends some ten miles inland, showing the part of the town with the works in question at one scale (480 feet to the inch), the rest of the area at another much smaller one (1400 feet to the

Maps and Fortifications

22 Isle of Sheppey, Kent, *c*.1572
A beautifully drawn plan intended probably to show the island's defensive potential. East is at the top. A note, bottom left, identifies anchorages and explains that Lord Cheney's lands are marked 'L:Ch:'. Queenborough Castle at the west end of the island, the cliffs on the north coast, beacons and windmills are all marked. The map-maker's monogram 'IM' is in the bottom right corner, but he has not been identified.
Public Record Office, MPF 240

23 Thames estuary, 1584
Part of a coastal chart that continues northwards as far as the Humber – a 'discription of the sandes' by Richard Poulter, mariner. Carefully drawn and coloured, and embellished with the royal arms, it was clearly meant for use in the study, not on board ship. Although many Tudor coastal maps have notes of depths and anchorages, this was to show where there was most danger of enemy landings, not to aid navigation.
British Library, Cotton MS. Augustus I.i.44

39

Maps and Fortifications

24 Hull, 1539
One of a series of plans drawn to show the state of coastal defences when attack was thought imminent and probably by the same hand as the contemporary plat of Scarborough (no.45). North is at the top, and on the east the town is defended only by the River Hull with a chain across its mouth. The figures beside the pictures of places on the left give their distance from Hull in local miles.
British Library, Cotton MS. Augustus I.i.83

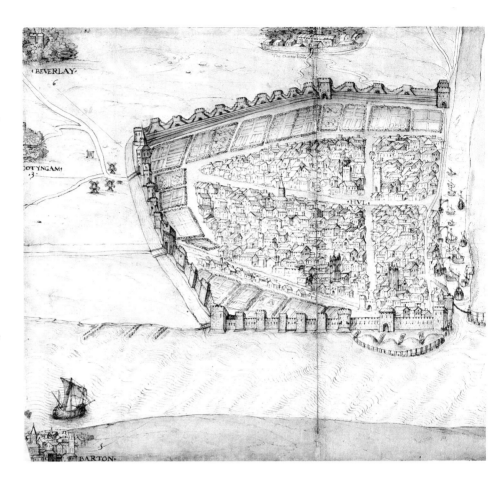

25 Hull, 1541–2
Over 6 feet long and probably drawn by John Rogers, this plan shows the scheme for fortifying the undefended east side of the town, the weakness demonstrated by the earlier plat (no.24) – west is at the top. It gives a very different view of the buildings fronting the River Hull and was probably drawn from life. On the nearer bank the projected wall and bastions are shown in plan. This and the other plans drawn for these works are the earliest known plans drawn to scale in England.
British Library, Cotton MS. Augustus I, Suppl. 4

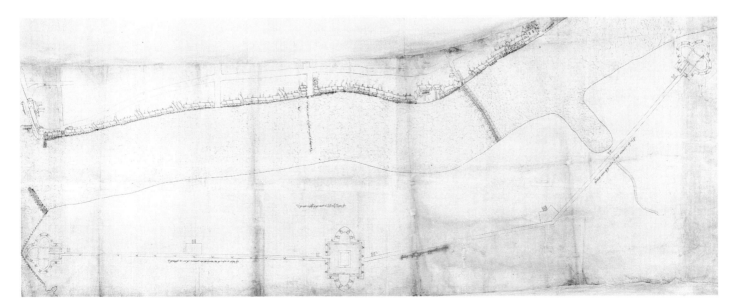

inch). A fine scale-map of the Isle of Sheppey, drawn about 1572 by 'IM', who has not been identified, distinguishes the fort at Swaleness by showing it in ground-plan, but covers the whole island in some detail, marking Lord Cheney's lands and particular farms. The plats of the engineers were far from strictly limited to the fortifications themselves, and they played a significant part in spreading the scale-map beyond the confines of the fort to the countryside at large.

fig. 22

Their mapping spread to coastal waters as well. We have seen how, at Dover Harbour, the engineering talent available to the Crown might be employed on other works besides fortifications. It was also used for nautical charts. In hydrography England lagged behind France and the Netherlands, and few charts of English coasts that give soundings were drawn before the end of the century. Those that survive were probably drawn as a guide to defence rather than for navigation, showing which points were most vulnerable to enemy landings. The earliest is of the Humber and the south Yorkshire coast, drawn about 1560. It is anonymous and we do not know why it was made. Others, however, were drawn by engineers in connection with royal works, among them the plan of Dover Harbour in 1581, and one of Portsmouth Harbour drawn about 1584 when the defences were being strengthened. Others too were produced by navigators, including William Borough, voyager to Russia, commander of a ship against the Armada and Controller of the Navy. But defence against enemy landings was a concern of the government long before engineers were used to introduce the precision of measured soundings into coastal charts – and it was a concern that led to the production of maps.

fig. 15

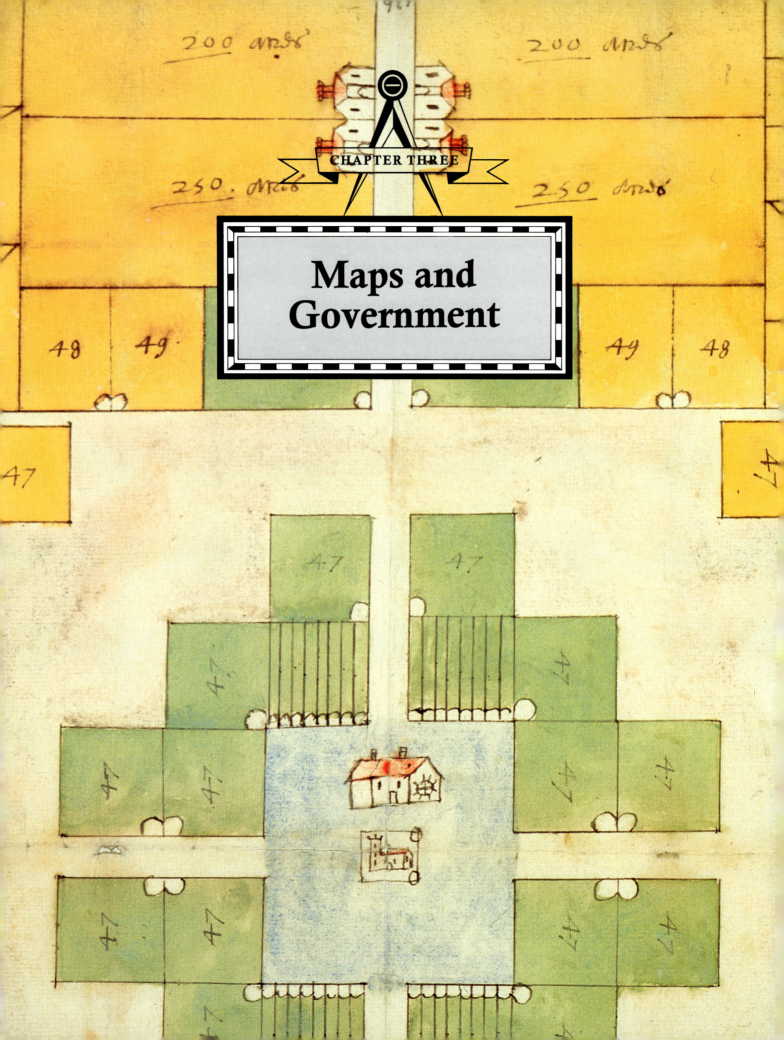

CHAPTER THREE

Maps and Government

Maps and Government

EVERY map had two parents, the map-maker who drew it and the customer or patron who commissioned it and paid for it. We have been looking at the use of plats in royal works from the point of view of those who made them – the engineers and other map-makers. But the king and his advisers played just as important a role: it was they who came to see that it was worth having plats made because they set out so clearly the problems to be solved in any scheme, the advantages and disadvantages of any proposed solution. Precisely how they were persuaded is still not clear. It may have been the work of the artists or engineers at court about 1530, or of advisers who brought back from the Continent, probably from Italy, news of these and other techniques used in fortification. But the idea could have come from south Germany, and a decisive part may have been played by humanists at court – Thomas More's *Utopia*, published at Louvain in 1516, was accompanied by a map of the imaginary isle, and in 1531 Thomas Elyot, in *The Boke named the Governour*, pointed out the value of maps to the shrewd ruler.

It used to be thought that several surviving maps of defences were drawn for Henry VIII early in his reign, and tentatively they were linked with the progress of the war with France in 1513–14 and with a proposed review of defences in 1519. However, research on individual maps has tended consistently to assign them to later dates. Thus, though it has been convincingly argued that a map of Brighton shows the French attack on the town in 1514, it is now thought that this was simply to show historically how vulnerable it was, that the map was drafted in 1539 and that the date it actually bears, July 1545, was when the existing copy was made. One map for which an early date still seems possible, and which has also been assigned to 1514 simply because of the risk of French attack at that time, is a picture-map in the form of a 25-foot long roll, showing the north Kent coast from Faversham to Margate. Along the other edge of the strip it shows the opposing shore: the other side of the inlet leading to Faversham, then the Isle of Sheppey – Leysdown-on-Sea is shown with its church – then a series of sandbanks parallel with the coast as far as Thanet. Opposite Shell Ness is a note beginning 'A goode rode for a C sayle of goode shepys' – that is, a good anchorage for 100 good ships. The two shorelines are shown with opposite horizons and the wording is one way up on one side, the other way on the other, so that there is no 'right way up' – a common technique in drawing picture-maps, suiting them for study by people on opposite sides of a table. It is crudely designed and painted and is clearly meant to be viewed at a distance rather than for close study of its detail, but it is peculiarly effective in presenting the coast's general character from the point of view of a mariner – or of an enemy fleet. But the suggestion that the map dates from 1514 is the merest guess, and despite its primitive appearance it may be much later – 1539 is at least as likely a date.

The fact is that we have no topographic map – nor any reference to one – that was undoubtedly drawn for the Crown before 1530. If the king and his ministers did make use of maps earlier for defensive and strategic planning – and there is no reason to suppose they did not – it was an occasional expedient, in contrast to the regular practice it was to become. Peter Barber has tellingly pointed to the difference between the defensive works in Picardy in the 1540s, for which whole series of maps and plans were made, and the fortification of Tournai in 1513–20, for which we know of none at all. It is interesting that for the festivities at Greenwich in 1527 on concluding peace with France Henry VIII's court painter, Hans Holbein, painted a 'plot' – apparently a mural picture-map – of the siege of Thérouanne in 1513. But, as far as we can tell at present, it may well have been simply through plats of individual fortifications that the government was introduced to the potential of cartography, having recognised its value first in this context. There is nothing to connect this innovation particularly with Thomas Cromwell or any other of the king's ministers – it seems rather to have been a device that was at once generally accepted.

But if the regular use of maps began in plans for individual fortifications and other royal works, they soon assumed a larger role in defence strategy. At the beginning of 1539 war seemed near: the pope had excommunicated Henry VIII, while Francis I of France and the Emperor Charles V had reached agreement in the treaty of Toledo. In February a survey of defences was ordered in all coastal counties, and this is thought to have produced a number of the surviving maps that show existing and proposed fortifications along the coast. All were picture-maps – it was to be another year or so before the scale-map appeared on the scene. Among them was the map of Brighton – perhaps also the map of the coast from Faversham to Margate. Others show Scarborough, the areas around Hull and Harwich, the coasts of Dorset and Somerset and, the most remarkable of the group, the south coasts of Cornwall and Devon, from Lands End to Exeter. This last is an outstanding monument of early-Tudor cartography. Nearly ten feet long, it is a detailed picture-map, showing the coastline with its ships, inlets, towns, villages, churches, bridges, parks and a host of defensive towers, some 'not made' or 'half made'. Notes beside the shore show where there were good or difficult landing-places, and the whole map was meant for close, detailed scrutiny, quite unlike the broad picture offered by the map of the north Kent coast. To the king and his advisers it showed, clearly and eloquently, in a way no written report could do, the points where the coast was most vulnerable and how they might best be defended. To us it gives a vivid panorama, realistic and picturesque, of a sizeable portion of Tudor England, showing how it appeared to its own rulers. Here we can show only portions of the map; it deserves, like the other maps in this group, full publication in facsimile and detailed analysis.

fig. 45
fig. 26

endpapers
fig. 32

Maps and Government

We have seen that when scale-maps of defence works were introduced in the 1540s some were soon drawn that took in tracts of country adjoining the fortifications. Before long it was scale-maps rather than picture-maps that were drawn for wider strategic, even political, planning. An early example was drawn in 1552 by Henry Bullock, who was subsequently Master Mason of the King's Works. It shows an area, some ten miles square, where the Scottish border was disputed at its western end, the so-called Debatable Lands. Four lines on the map mark the border as separately defined by the English commissioners, by the French ambassador, by the Scottish offer and then 'the last and fynal Lyne' agreed on 24 September 1552. The threat of the Armada in 1588 produced not only plans of individual works, as at Plymouth and Portsmouth, but strategic plans of defence. One, by Robert Adams, covers the lower reaches of the Thames, from London to Tilbury. It shows successive batteries along the river with the area that each covered with its guns, and two barriers that were projected at Tilbury and Blackwall. Another, showing the coasts of Devon and Cornwall 'as they were to bee fortyfied in 1588 against the landing of any Enemy' must be a copy, made not much later, of a map drawn up at the time and shows not only fortifications but bodies of troops as well. By then, maps –

fig. 29

fig. 33

26 Portland and Weymouth, 1539–40
Part of a map of the Dorset coast, one of a group produced to show the state of coastal defences when attack threatened in 1539. West is at the top, and Melcombe Regis is in the foreground. Radipole Lake, between Melcombe and Weymouth, is prominent, but Portland, on the left, is underemphasised. Five beacons and three coastal crosses can be seen.
British Library, Cotton MS. Augustus I.i.31,33

45

Maps and Government

scale-maps – were a normal instrument of defensive and military planning at every level.

In Henry VIII's reign no individual minister can be particularly linked with the introduction of maps to the king's government, and throughout the reign of Elizabeth I the use of maps must have been familiar to all her advisers. However, it is probably right to attach special importance to the interest in maps shown by William Cecil, who became Lord Burghley in 1571 and who was Secretary of State in 1550–3 and 1558–72, then Lord High Treasurer from 1572 until his death in 1598. It is natural to do this from the evidence of surviving maps, though these may simply reflect his natural acquisitiveness and the survival of his personal and official archives more or less intact. We have some half-dozen sketch-maps in Burghley's own – very distinctive – hand. They are oddly varied, and include one of the Debatable Land in Liddesdale, one of the River Lea, east of London, from Hackney to Bow Bridge, and one showing places on the route from London and Windsor in the south to Bowes Castle and Brancepeth in north Yorkshire and County Durham. Far more often, however, we see Burghley's hand on maps by other map-makers, for he was an inveterate annotator of documents in general and of maps in particular. Many of his notes, as we would expect, refer to defence. Others show the same interest in communications – routes and staging posts – as his own sketch-map of places between London and the north-east. Far more, however, are names of local

fig. 34

27 Brighton, Sussex, 1539–40
A vivid picture of the French attack on the town in 1514, but probably a copy made in 1545 of one of the maps drawn to show the state of coastal defences when new threats arose in 1539 – it demonstrates Brighton's vulnerability all too clearly. Notes on the map tell how the big ships anchored offshore, while the landings were made from galleys. The beacons are alight, and a local defence force is advancing down the road.
British Library, Cotton MS. Augustus I.i.18

28 Whitstable and Seasalter, Kent, early sixteenth century
A short section of a roll 25 feet long that shows the north Kent coast from Faversham to Margate. South is at the top and at the foot are the sands of the Oaze. The style suggests a date early in the century, but it may possibly be one of the coastal maps drawn in 1539–40.
British Library, Cotton Roll XIII.12

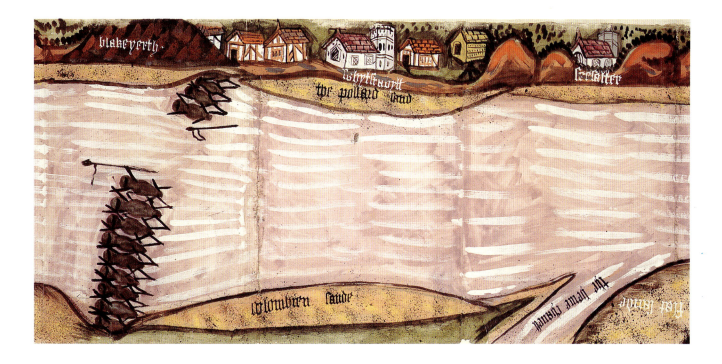

47

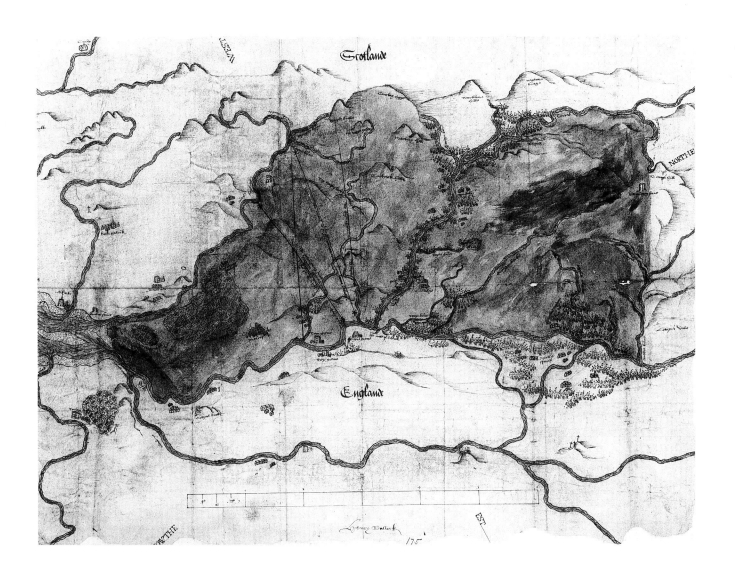

gentry – a list of 'Names of y^e principall lordships in y^e Middle march with the Lordes names' on a map of Northumberland for instance, or names of Catholic recusants entered by their homes on the map. On one map he entered lines of doggerel on some Essex houses:

> Heyghfeld fayre and fatt,
> Barndon Park better than that,
> Coppledon beares a crown,
> Copthall best of all.

The maps Burghley annotated were his own – his collection of maps is itself evidence of the interest he took in them. Some older maps he clearly acquired

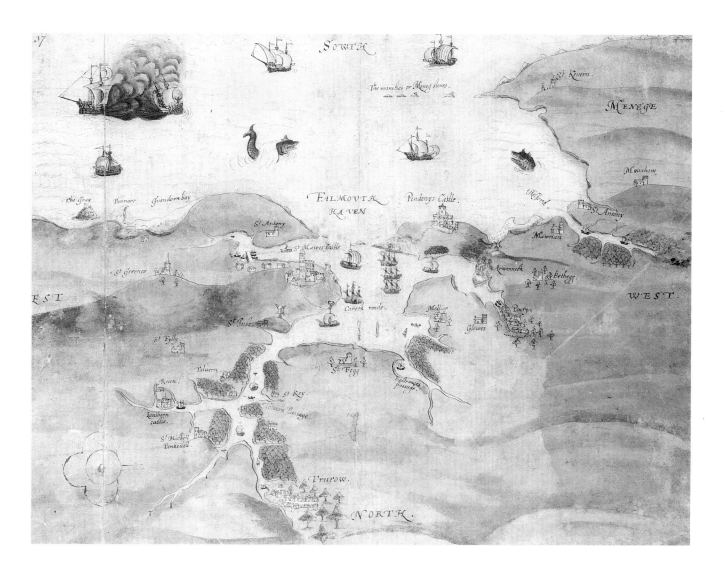

29 opposite The Anglo-Scottish border in Cumberland, 1552

An early example of a scale-map drawn for the Crown to show a broad tract of country, not just a single fortification or other proposed work. It is oriented west by north-west, with 'WEST' – and 'Scotlande' – written at the top. Across the middle of the map are Liddel Water and the River Esk, which meets the Lyne and the Eden on the left. In the centre, south of Canonbie, ruled lines show four different alignments of the Anglo-Scottish border. The map-maker, Henry Bullock, is named at the bottom, below the scale-bar.

Public Record Office, MPF 257

30 Falmouth Harbour, Cornwall, mid-sixteenth century

Showing the layout of the harbour and its surrounds, the map was drawn after the completion of St Mawes and Pendennis Castles in the 1540s. While maps for defence planning were increasingly drawn to scale from measured survey, picture-maps in older style long continued to be drawn. However, the clover-leaf pattern, bottom left, is an accurate ground-plan of St Mawes Castle by an annotator perhaps exasperated by its fanciful picture on the map. The naval battle, top left, is simply an embellishment.

British Library, Cotton MS. Augustus I.i.37

Maps and Government

31 Mount's Bay, Cornwall, mid-sixteenth century

This was presumably drawn in connection with defence, but on what occasion is not known. In 1539, however, Mount's Bay was listed among places on the south coast needing fortification. South is at the top. Penzance is lower right and St Michael's Mount is in the centre.

British Library, Cotton MS. Augustus I.i.34

32 The River Dart and Tor Bay, Devon 1539–40

A section of the panoramic map, nearly 10 feet long, showing the whole coast from Lands Ends to Exeter, that was drawn to plan coastal defence in the south-west. Kingswear and Totnes (with bridge) appear as towns, but Paignton only as a name between two towers. The long notes beside the shore describe possible landing-places and list distances between points along the coast.

British Library, Cotton MS. Augustus I.i.35,36,38,39

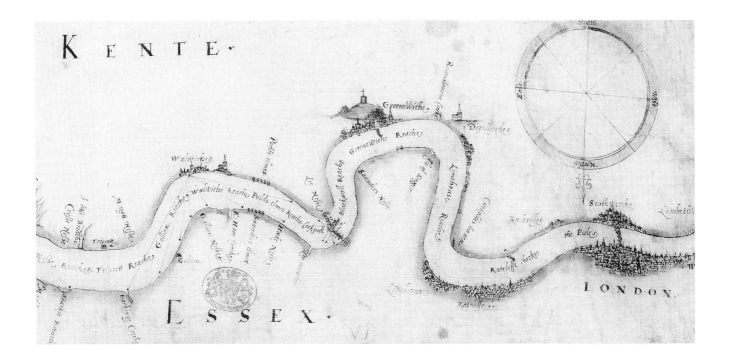

33 Defences on the lower Thames, 1588
Part of a map – of which more than one contemporary copy survives – drawn by Robert Adams to show the defences on the Thames below London at the time of the Armada. This portion shows the reaches covered by the battery at Woolwich and the effect of the fortified barrage at Blackwall.
British Library, KTC VI.17

from official records, such as a plan of fortifications at Ambleteuse in 1546–7. Another, of the fort at St Mary in the Scilly Isles, was probably drawn in 1551 and may have passed to Burghley when Robert Adams was sent to the Scillies in 1591 to prepare defence works. He seems to have been prepared to keep any map that came his way, without great regard for its quality. Many maps, printed and manuscript, he had bound up into what were in effect personally compiled atlases which he used as works of reference, the more useful to him for their many annotations. The oldest of these, containing maps by Lawrence Nowell of Britain, Ireland and Sicily, was of notebook size and Burghley is believed to have kept it always with him. The others that survive are large books. One contains some fifty maps for England and Wales, another is of foreign countries, the two presumably being meant as a pair for reference on either domestic or foreign affairs. Neither covered Ireland, but a list of Burghley's books in 1598 included two atlases of Irish maps which have also been identified with surviving volumes. The nucleus of his atlas for England and Wales was a set of proofs of the county maps by Christopher Saxton, published between 1574 and 1578, and this at once raises the question how far the government – or Burghley – was behind this and other schemes to produce detailed general maps covering the whole country.

The earliest such scheme seems to have been a project of John Rudd, a cleric born in Yorkshire about 1498, who bowed before every change in the wind of

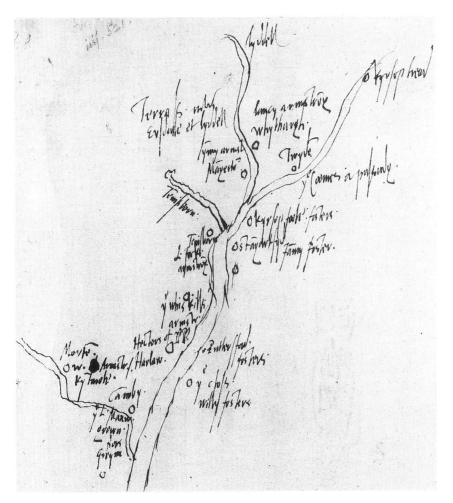

34 Liddesdale, by William Cecil, *c*.1561
Cecil, who became Lord Burghley in 1571, took a keen interest in maps, but only a few sketch-maps survive in his own, easily recognisable, hand. Liddesdale was at the heart of the Debatable Land, then at issue between England and Scotland, and Cecil has noted the names of the principal houses and their owners along Liddel Water and its tributaries.
Public Record Office, SP 59/5, f.44

religious policy – under Edward VI's protestantism he married, under Mary I's catholicism he publicly repented and renounced his wife, under Elizabeth I he married her again. He had long been interested in maps and in 1561 he claimed to have been to much trouble to make a plat of England, and proposed to spend two years travelling throughout the country in order to perfect it. He was then a prebendary of Durham, and the queen ordered the cathedral chapter to continue to pay him during his absence – clear evidence of the government's interest in an undertaking of which in fact we know little. It produced no map that survives, unless Rudd's work contributed to the map of the British Isles that Gerard Mercator, working at Duisburg, published in 1564. More significant, perhaps, is the fact that Christopher Saxton, another but much younger Yorkshireman, seems to have been working for Rudd in 1570, when he signed a receipt for some money for the use of John Rudd 'my master'.

In 1563, however, a new scheme was put forward. Lawrence Nowell – for long confused with his cousin and namesake, the Dean of Lichfield – was a member of William Cecil's household. He was a traveller, Anglo-Saxon scholar and cartographer, and he now proposed to survey and map the whole kingdom. This produced the small-scale map in Burghley's earliest atlas but nothing more, and the idea must have been abandoned. Other surviving maps by Nowell include one of Scotland, drawn rather more accurately than Mercator's map of 1564, and one of England with place-names in Anglo-Saxon lettering – but no large-scale maps of smaller areas of the country. Whatever royal or official enthusiasm there may have been for such a project seems now to have waned, and it was several years before any further step was taken.

fig. 35

fig. 12

It was then not Burghley or the government that commissioned Christopher Saxton's county maps but Thomas Seckford. Seckford, born in 1515, came from Woodbridge in Suffolk. He was a well-to-do barrister, and held the office of surveyor of the Court of Wards and Liveries. He must have paid for Saxton's work, for his coat of arms, with a motto of his choice, appears on every map. But there seems no doubt that from the start the project had the approval and support of Burghley, as well as of the queen and probably others of her ministers. A later record suggests that Saxton had letters from the queen supporting his work in 1573; certainly in 1574, the date that appears on the earliest of the maps, the queen granted him a lease of lands in Suffolk because of the 'grand charges and expenses lately had and sustained in the survey of divers parts of England', two years later a letter from the Privy Council ordered Justices of the Peace and other officials in Wales to give him every assistance in his survey, and in 1577 the queen gave Saxton the exclusive right for ten years to publish his maps. Burghley, himself master of the Court of Wards and Liveries since 1561, had long known Seckford, and his possession of proofs of the maps shows his interest and involvement.

Christopher Saxton was born about 1542–4. He came from Dunningley in Yorkshire, three miles from Dewsbury, where John Rudd was vicar from 1554 to 1570 – Rudd held the vicarage and other livings along with his Durham prebend. We know nothing for certain of Saxton's early life before he produced his county maps, apart from the tantalising reference to Rudd as his master in 1570, and we can only assume that it was from Rudd that he learned cartography and surveying, and probably through him that he was recommended to Seckford. Nor do we have any map or written survey by Saxton earlier than his 34 county maps, which bear dates between 1574 and 1578 (one is undated). The maps of the Welsh counties are all dated 1577 or 1578, but the queen's letter ordering local help for him there was issued only in 1576. This suggests that there was no long interval between Saxton's survey of any particular county and the date engraved on its map, and thus that work on the English counties

35 Scotland, by Lawrence Nowell, mid-sixteenth century

Comparison with the anonymous map of 1534–46 (no.2) shows how rapidly the mapping of Scotland improved in this period. It shows too Nowell's quality as a cartographer – his map is significantly better than Gerard Mercator's of 1564. He worked partly from existing maps, partly from his own surveys. It is not known why his project of a detailed map of England was abandoned.

British Library, Cotton MS. Domitian A.xviii, ff.98v–99

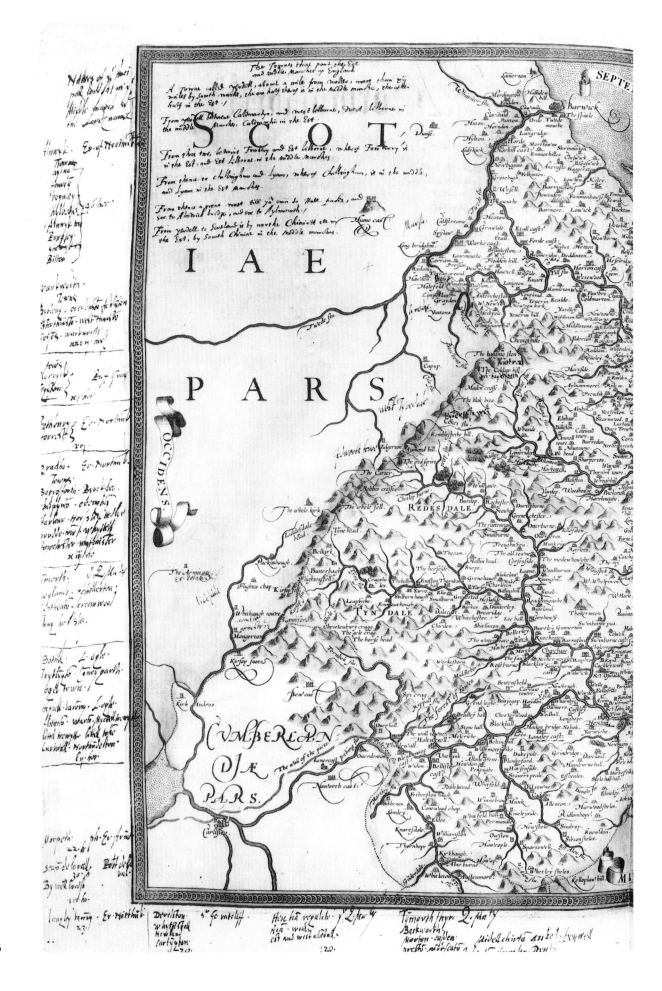

Maps and Government

36 Annotations by Lord Burghley on Saxton's map of Northumberland, late sixteenth century

One of the maps in Burghley's personal atlas of England and Wales, which has a proof set of Saxton's county maps as its nucleus. He has added notes to the map itself – as 'West Tyvedale', West Teviotdale, at the end of 'SCOTIAE PARS' – and in the margins, where top left are 'Names of ye principall lordships in ye Middle march with the Lordes names'. In the top left corner of the map, notes of places along the Anglo-Scottish border are in a different hand, perhaps a secretary's.
British Library, Royal MS. 18.D.iii, ff. 71v–72

37 Kent, Middlesex, Surrey and Sussex, by Christopher Saxton, 1575

One of the earliest of Saxton's county maps. The arms of the queen (top centre) and of Thomas Seckford (bottom right) and the scale-bar with dividers (centre right) and date (top right) occur on nearly all the maps, but the Latin notes of the number of towns and parishes do not. The strapwork and the allegorical figures are integral parts of a map that was designed for display.
British Library, Royal MS. 18.D.iii, ff. 23v–24

57

Maps and Government

38 Monmouthshire, by Christopher Saxton, 1577
At nearly a half-inch to the mile, this is drawn at more than twice the scale of Saxton's map of the south-eastern counties. Like no.36, this is the copy of the map in Lord Burghley's atlas, and a few of his annotations can be seen, such as 'Herb't', Herbert, just east of Abergavenny.
British Library, Royal MS. 18.D.iii, ff.100v–101

39 Isle of Wight, by John Norden, 1595
Drawn in a separate panel beside the map of Hampshire in a manuscript containing Norden's 'Chorographicall discription' of Middlesex, Essex, Surrey, Sussex, Hampshire and the Channel Islands. The account of Hampshire was not published. The symbols correspond to the key on the main map but are not applied comprehensively – a tower with a steeple is a parish, houses with a crescent above are a market town, a squat tower is a castle if it has a flag, otherwise a nobleman's house.
British Library, Additional MS. 31853, f.24

began only a year or so before 1574. Presumably each map was printed and published as soon as it was ready – though this cannot be taken as proved – but when the series was complete the maps were all issued together in an atlas which also contained a portrait of the queen as frontispiece, a list of the maps, and a general map of England and Wales, all dated 1579.

Most of Saxton's maps cover a single county, but some groups of counties appear together – Kent, Surrey, Sussex and Middlesex, for instance, or Montgomery and Merioneth. This was not to achieve uniformity of scale, for this varies from one map to another; several are drawn at almost five miles to the inch, but the scale of one, Monmouthshire, is nearly twice as large. We can only guess how Saxton managed to produce the maps with such speed and efficiency or, indeed, what method he used in surveying. It seems most likely that he constructed a framework for each map by triangulation – those ordered to help him in Wales were to 'see him conducted unto any towre Castle highe place or hill to view that countrey' – and then inserted details within this framework by measuring routes or by simple sketching. William Ravenhill has made the interesting suggestion that his triangulation may have made use of the network of alarm beacons. He may have drawn heavily on existing maps, whether sketched or surveyed, but we have yet to discover a direct connection between

fig. 37

fig. 38

fig. 13

Maps and Government

Maps and Government

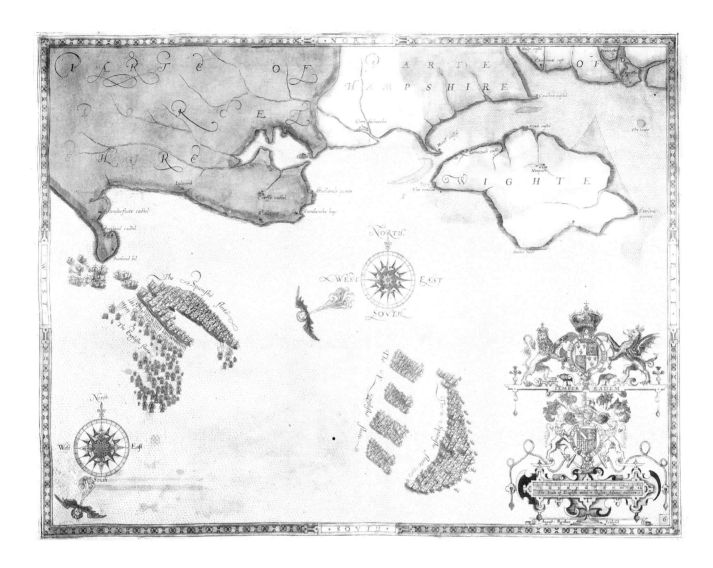

40 The Spanish Armada off Portland and the Isle of Wight, 1588
One of a set of eleven maps showing each stage of the arrival and defeat of the Spanish Armada. Engraved from drawings by Robert Adams, they illustrate an account of the Armada published in Italy in 1590. This, the sixth map, shows two successive positions in the operations of 24 July 1588.
British Library, Maps C.7.c.1

one of Saxton's maps and any earlier map of the same area. Despite occasional lapses – the outline, or direction, of the Cornish peninsula is the most notable – the maps are impressively accurate and detailed.

They are in fact very good maps indeed, and they at once put England, which had fallen behind in cartographic development, at the forefront of European mapping. There was, however, room for improvement – the maps did not show roads nor, in many cases, how the counties were divided into hundreds or wapentakes – and before the end of Elizabeth's reign two further schemes for maps of the counties were begun. Despite some degree of government backing neither progressed far. The first was by John Norden, surveyor, who planned a description and map of each county under the general title of *Speculum Britan-*

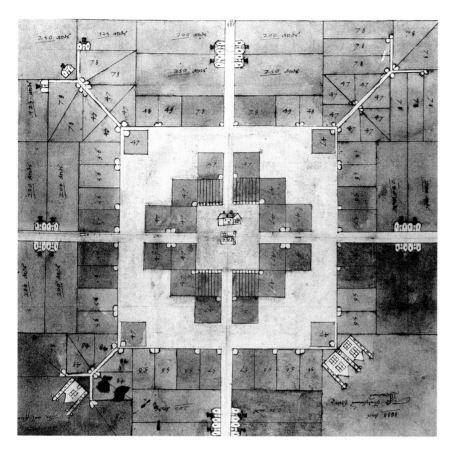

41 Scheme for a new lordship in Munster, 1586

An idealised plan for a settlement, explained in notes beside it. Thus 'under the figures of 1000 is the gentlemans Scite and demeanes' (bottom right), and the parcels of 200 acres (centre of each side) are freehold. In the centre are the parish church and 'the parsonage & vicarage howses', with what may be a mill-wheel.

Public Record Office, MPF 305

niae, the mirror of Britain. Middlesex was published in 1593, Hertfordshire in 1598, and several other counties – text and map – were completed and survive in manuscript but were not published. They are notable maps: unlike Saxton's they show roads, and most include a key to conventional signs for towns, villages, castles, mills and so on, and are the earliest English regional maps to do so. Norden may have lost the official support which gave him, like Saxton, local access and help in making his survey as well as the rights of publication. Certainly the scheme was abandoned, but a new project for a map of each county was soon begun by William Smith, antiquary and now herald – he became Rougedragon Pursuivant in 1597, a token of royal favour perhaps connected with his county maps. These, again with keys to their symbols, were very finely engraved – in the Netherlands – and are dated 1602 and 1603, but they cover only twelve counties. It was left to John Speed, a few years later, to produce the first full set of county maps comparable to Saxton's.

Most of Speed's county maps were no more than revised versions of Saxton's, which remained the basis of nearly all county maps – and thus of nearly

Maps and Government

Maps and Government

42 Lough Neagh, Counties Armagh and Tyrone, 1600–3

Part of a map drawn at the time of the earl of Tyrone's rebellion – top right are the arms of the queen and of Lord Mountjoy, Lord Deputy of Ireland. The map identifies the family dominant in each area and gives a lively impression of the rivers, hills and woods, but shows also, in the many forts and roofless buildings, that it was a country suffering in war.

Public Record Office, MPF 36

43 Siege of Enniskillen Castle, County Fermanagh, 1594

'Made and dun by John Thomas Solder', like an equally spirited picture of the battle at Belleek the previous year – nothing more is known of the artist. It shows how the castle was taken, using cannon in trenches, a 'greate bote' – with 67 men on board – rowed round and anchored by the wall to breach it, and scaling ladders on the farther side of the castle. In 'Governor Dowdalls Campe', bottom left, are 'The howse of munition' and the heads of three rebels – two of them bearded – on posts.

British Library, Cotton MS. Augustus I.ii.39

63

all regional cartography in England and Wales – until the large-scale private surveys of individual counties in the eighteenth century. But it is not just for their longevity, nor even for their quality, that Saxton's county maps are so important. It is rather because they succeeded in introducing cartography to the literate public at large. They were maps as we understand them, drawn to consistent scale and with no trace of the bird's-eye view or picture-map beyond the pictorial form of some conventional signs. Hitherto maps of this sort, such as the scale-plans of projected defences, were familiar only to limited groups – engineers, navigators, administrators, antiquaries – as a particular technique or tool of their craft. Other people had become used only to maps that were, superficially at least, simply pictures. The only local or regional maps published earlier than Saxton's – of London and a few other towns – were all in this form. Saxton's county maps, however, achieved a crucial break-through in general understanding. That they sold well is likely from the many copies surviving today, and we can see in other ways how these maps entered into popular consciousness. Tapestries woven at workshops in the Midlands bore huge versions of Saxton's maps, and at the other extreme of size playing-cards were produced in 1590 with a tiny county map on each, taking advantage of the

44 Haulbowline Fort in Cork Harbour, late sixteenth century

The fort is built on a small island, but a note on the left explains that 'ye Iland is much bigger than here the figure shewes' – the scale applies only to the fort itself. The unknown map-maker drew similar plats of other forts in Ireland.

British Library, Cotton MS. Augustus I.ii.36

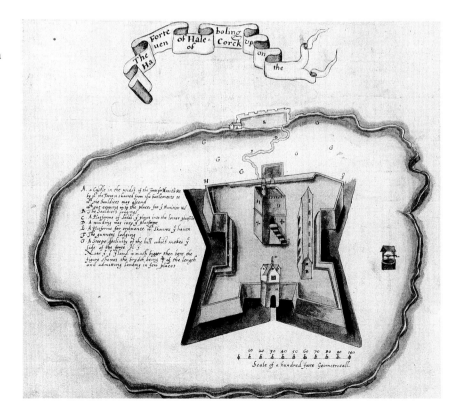

happy chance that there were 52 English and Welsh counties, one for each card in the pack. Also in 1590 an English translation was published of Petruccio Ubaldini's Italian narrative of the defeat of the Spanish Armada two years earlier, and with it was a magnificent set of eleven engravings showing successive stages of operations. Drawn by Robert Adams, they show the English and Spanish fleets pictorially, but sailing on scale-maps of the English Channel. It was now no longer royal ministers alone who saw strategy and defence in terms of maps.

fig. 40

Certainly Saxton's maps were put to immediate use in every level of government, and Burghley's were not the only copies of the maps to be annotated with lists of local Justices of the Peace. George Owen, whose *Description of Pembrokeshire* was published in 1603, thought that one reason why his county was specially heavily burdened by the Crown was that Saxton had given it a map all to itself, whereas neighbouring counties were drawn on a smaller scale and combined on a single sheet – this had led ministers, overlooking the difference of scale, to suppose that Pembrokeshire was larger than it really was. By the end of the century, indeed, maps – nearly always scale-maps – were being used for the everyday work of government in every sphere, to an extent unknown even 50 years earlier. This can be illustrated from activity in the latter part of the century in one clearly defined area, Ireland. Hand in hand with the extension of English rule went the mapping of successive provinces by cartographers employed by the Crown – Robert Lythe, the two John Brownes, uncle and nephew, and Richard Bartlett. Then, whenever practicable, maps were used in solving problems of security and settlement, as we see not only from plans of forts and maps of land division but from – for example – a map of two new counties projected in 1579, a model plan of a new lordship in 1586, a map to show how 11,000 troops could be garrisoned in Ulster in 1598. Meanwhile private cartography produced not only some spirited pictorial maps of battles and sieges but also Ireland's earliest estate map, dated 1598, of Sir Walter Raleigh's lands at Mogeely, County Cork. The chronology of mapping in sixteenth-century Ireland differed significantly from what happened in England, but it shows even more clearly how the government's use of maps led to general awareness of their value.

fig. 44
fig. 41

fig. 43

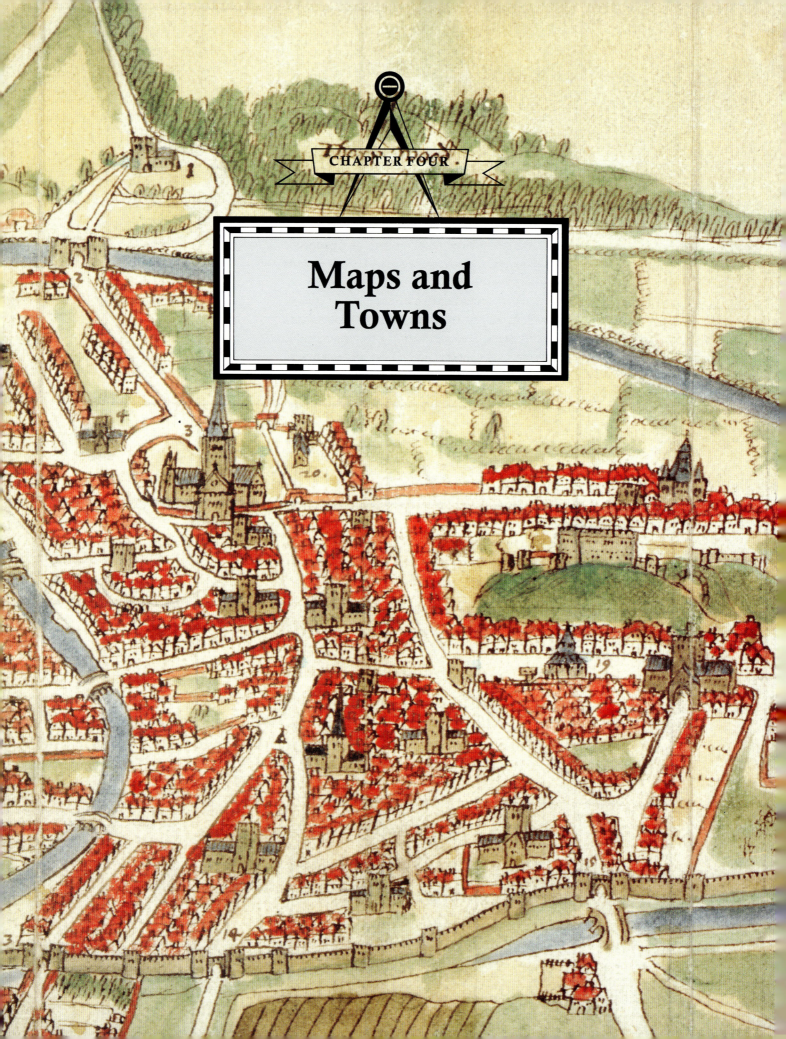

CHAPTER FOUR

Maps and Towns

Maps and Towns

OTHER chapters in this book describe the use of maps in contexts where we can see a connected theme of development. In each case the maps were drawn to meet a particular need and they increased in number and changed in form as the sixteenth century progressed. Tudor maps of towns are not like this. It was only late in the century that a coherent English tradition of town plans and views began. Earlier, most towns on maps appear only incidentally – the maps themselves were drawn primarily for some other purpose. Thus a plat of a town's fortifications seldom leaves the town itself blank, but the plat was not drawn to provide a picture of streets, churches and houses. A map of coastal defences may show towns in some detail, though only their walls, harbours and other landing-places are really relevant to it. But although the town itself may sometimes appear almost by chance, maps of towns are among the most interesting of Tudor maps – and are certainly among the most picturesque and informative, giving us an extraordinarily full and lively view of what these towns looked like, how they appeared in the surrounding landscape. In looking at them we are reminded that towns in Tudor England were not simply centres of trade, industry and in some cases wealth – they were also objects of pride to their inhabitants. There were fewer of them than there are today, and although most were much smaller than modern towns there was a sharp distinction between town and country, despite buildings in the outskirts and the occasional straggling suburb.

Our only medieval map of an entire English town is of Bristol, drawn about 1480 to illustrate the local chronicle that was written by Robert Ricart, the town clerk. It is a bird's-eye view. The four gates of the town and the High Cross in the centre are drawn unnaturally large and are named, and churches and some other buildings have recognisable outlines, but most of the town is shown as a mass of conventionally drawn roofs and frontages. It is conceivable that this owed its inspiration to the views of cities that were being drawn in contemporary Italy – Bristol was a port and such contact was not impossible. But we are on surer ground in seeing Italian tradition behind the bird's-eye views of towns that we find on sixteenth-century maps of fortifications and coastal defences. Many are drawn more or less conventionally, like the towns of south Devon and Cornwall that appear on the coastal map of 1539–40. But some achieve greater realism. A map of Hull which may stem from the same survey of defences in 1539, serving as a starting-point for the works put in hand there the following year, shows the town within its walls and beyond it the Carthusian Priory, the houses at Drypool and, in the distance, other neighbouring towns and villages. Within the town the streets and houses are not accurately drawn, though some buildings can be identified – among them two churches, two friaries and the town gaol – and beside the walls we see gardens in areas not built up. In concept, though not in style, it thus resembles the Bristol map of 60

endpapers
fig. 32

fig. 24

67

Maps and Towns

years earlier, but the town walls and the chain and cannon at the entrance to the River Hull, the defences that were the map's prime purpose, are drawn in exact detail.

In 1541, however, maps not just of town defences but of towns themselves were being drawn for the government, following an act which restricted rights of sanctuary to the precincts of cathedrals, parish churches and hospitals, abolishing the right in some wider areas where it had been traditional. The mayor of Norwich was ordered to have a map drawn on parchment to show the areas of sanctuary within the city; the map survives with the king's letter attached to it. The city of York's records enter payment for a similar map there – drawn, curiously, by 'the armet (*i.e. hermit*) of the Kyngs Manour' – and this has been convincingly identified with a map of which only two large fragments survive.

fig. 46

45 Scarborough, *c*.1540
Probably one of the maps of coastal defences produced under threat of attack in 1539 and by the same artist as the map of Hull (no.24). Like the map of Hull it shows how a plat of the defences may incorporate a full picture of a town. Such a map may be a useful guide to the general appearance of the town and of its principal buildings, but cannot be relied on in its detail of individual streets and houses.
British Library, Cotton MS. Augustus I.ii.1

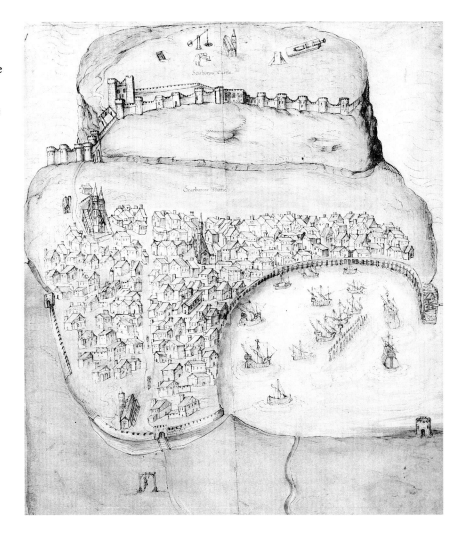

Maps and Towns

A map of Southwark is thought to have been drawn on the same occasion. This is the first time we see the Crown ordering other authorities to have maps made, or, indeed itself using maps for administration other than in planning defences. The three surviving maps – we may assume that more were drawn – illustrate clearly the contemporary state of the map-maker's art. None is drawn to scale. The Southwark map shows the plan of streets simply sketched, with inns, churches and a few other features, including pillories, drawn roughly in elevation. The York map is more carefully drawn but is more diagrammatic – the River Ouse is shown as a straight line and streets are set out in a rectangular pattern, with only churches marked by a conventional sign, a tower with a spire and cross. The Norwich map also shows only selected features along the pattern of streets, but these appear in realistic bird's-eye view and include the bridges, market place with cross and, again, the pillory.

fig. 49 Quite different from all these is a plan of Portsmouth drawn in 1545. It is one of the most remarkable of all maps from Tudor England – and the fact that to our eyes it seems utterly unremarkable shows what an extraordinary production it is. It was drawn to set out improved defences for the town after attack by the French. Drawn to scale – 'Thys plat ys In every Inche on C fote' (that is, 1 inch to 100 feet) – it shows the walls and bastions, the streets and the buildings along the streets all entirely in outline ground-plan, without a single pictorial

46 Norwich, 1541
Part of a map showing areas of sanctuary in the city – the king ordered the mayor to have it drawn following limitations on rights of sanctuary. The map is badly worn and damaged, but visible here are the market-place with the 'Crosse in the market' and, to the right, 'saynt petar chirche', St Peters Mancroft.
Public Record Office, MPI 221

Maps and Towns

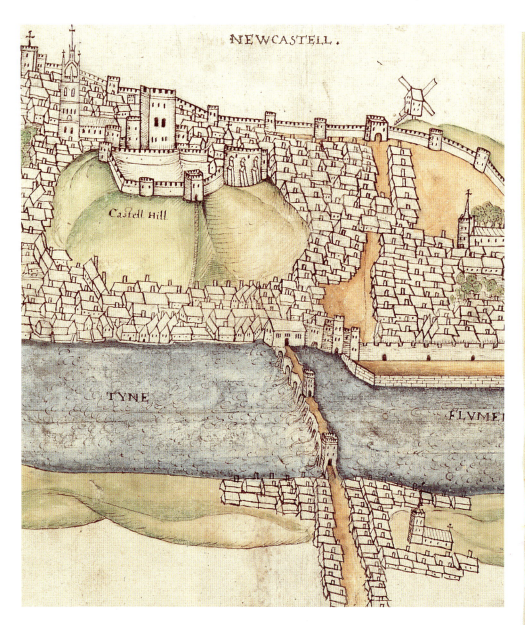

47 Newcastle-upon-Tyne, mid-sixteenth century

The central portion of a more extensive plat, perhaps drawn by Gian Tomasso Scala in 1545, when large-scale works of fortification were under way at Tynemouth, 8 miles down the river. The town is viewed from the south, with Gateshead in the foreground.

British Library, Cotton MS. Augustus I.ii.4

48 OPPOSITE Shrewsbury, late sixteenth century

One of the manuscript maps supplementing Saxton's county maps in Lord Burghley's own atlas of England and Wales. West is at the top, and 'ye welsh gate' by the bridge across the Severn (top centre) is in Burghley's hand. The author and origin of this map are not known.

British Library, Royal MS. 18.D.iii, ff. 89v–90

Maps and Towns

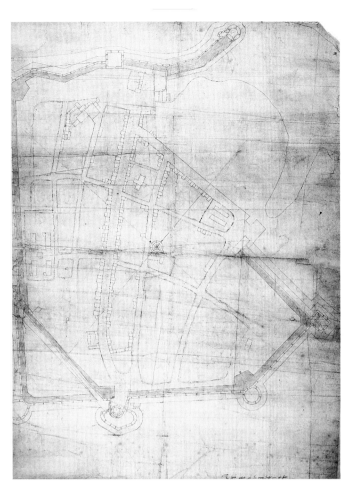

49 Portsmouth, 1545

An undistinguished-looking map, but very remarkable in the context of cartographic history. It is the earliest map of any town in Britain to be drawn entirely as a scale ground-plan – no feature is shown pictorially. There are in fact few parallels anywhere in Europe at this date. Drawn to set out proposals for improving the town's defences after attack by the French in July 1545, it includes alternative plans for the south-east corner (bottom left). Some later proposals were added in pencil, probably in 1546.

British Library, Cotton MS. Augustus I.i.81

50 Carlisle, c.1563

A scale-plan with pictures superimposed, drawn perhaps to show the need for work on the walls – stretches on both sides of the town appear badly damaged. The pointing finger beside the name of each cardinal point – 'Southe' is just above the scale-bar – is an unusual feature. In the market-place the pillory is prominently drawn, as so often in Tudor town plans.

British Library, Cotton MS. Augustus I.i.13

feature. The frontages of the houses may have been measured or simply sketched in, as their back walls almost certainly were, or even entered conventionally to show simply that there were houses along the streets – but this is beside the point. The idea of representing a whole town solely by the scale outlines of its structures and boundaries is cartographically far advanced. It has no parallel in sixteenth-century England, and few anywhere in Europe in the first half of the century. Probably more comprehensible, and certainly more acceptable to the sixteenth century, was an elaboration of this, taking just such an outline ground-plan as a base and imposing on it pictures of all the buildings and other features, a technique that William Cuningham illustrated with a plan

51 London, 1553–9

No copy survives of this first map of London – only two of the twenty engraved copper plates from which it was printed. One, shown here from a recent impression, covers the north-east corner of the city and the adjacent suburb – the site of Liverpool Street station is about its centre. On the left are tenter-frames for drying cloth after fulling; on top of the two city gates the heads of executed criminals are stuck on poles.
Museum of London

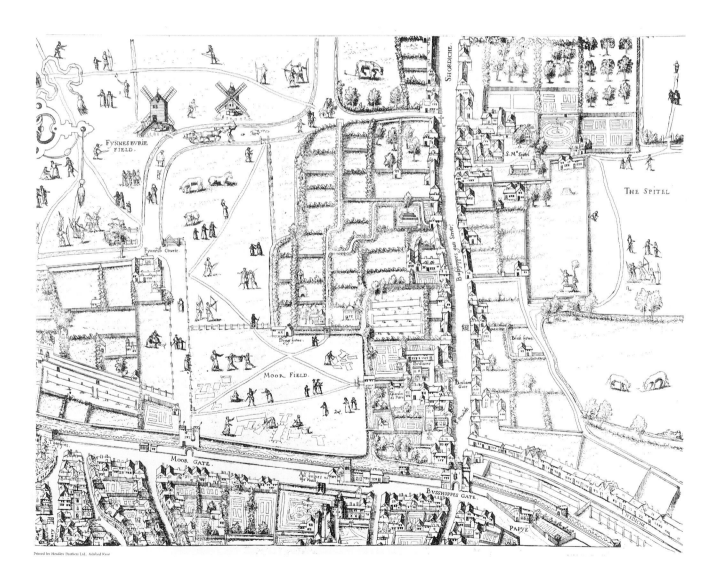

of Norwich in his book *The Cosmographical Glasse*, published in 1559. An impressive early example of this is a plan of Carlisle that was drawn probably in 1563 to show the poor state of repair of the town walls. At first sight it is a bird's-eye view, drawn in simple perspective, but in fact it is a scale-plan – there is a scale-bar – with each building and other feature drawn individually: they are not drawn smaller if they are further away from the imagined position of the draughtsman.

This technique of imposing pictures on a scale ground-plan is used for the earliest printed maps of London. Of the oldest, compiled between 1553 and 1559, no copies survive and its existence was discovered only when two of the original engraved copper plates from which it was printed came to light in 1955 and 1962 – about the end of the sixteenth century Flemish artists had used the backs as panels for oil paintings. They are only two of probably 20 plates that were used to print what must have been an enormous map, some 5 feet deep and 8 feet long, but they happen to cover adjacent sections and show that the map was the source for two other printed maps of London of which copies do survive. One of these is a much reduced version that appears in the monumental collection of city plans and views entitled *Civitates Orbis Terrarum*, cities of the world, that Georg Braun and Frans Hogenberg published at Cologne in a series of volumes from 1572 onwards. The other is less reduced – it measures just over 2 feet by 6 feet – and is printed from eight woodblocks, but of this map, which has been attributed for no good reason to Ralph Agas, we have only three copies, all printed in the seventeenth century after there had been some alterations to the blocks. However, from this miscellany of evidence we can get a fairly clear notion of the original map of the 1550s – and very impressive it must have been. Ostensibly it offered a picture, drawn from life, of every house in the city, and of many less permanent features as well. In fact, although principal buildings and much else besides were certainly drawn from life, the houses along the streets and lanes and even some of the churches have probably been entered conventionally. All the same, we gain an extraordinarily detailed and reasonably accurate impression of the close-packed London of the mid-sixteenth century, its suburbs beyond the walls and the adjacent Westminster and Southwark, administratively separate but all parts of a single conurbation.

Although we have no copies of the map of the 1550s, the two engraved plates are worn through use and many maps must have been printed from them – the survival of very large maps on paper is bound to be chancy. Like its two derivatives it probably had at the top the arms of the sovereign and of the City of London, and a decorative panel with some more or less grandiloquent title. It was a statement – a widely accepted statement – of London's size, power and prosperity, not a guide to its maze of streets. But London was not the only English city to evoke local pride, and map-makers followed its example else-

Maps and Towns

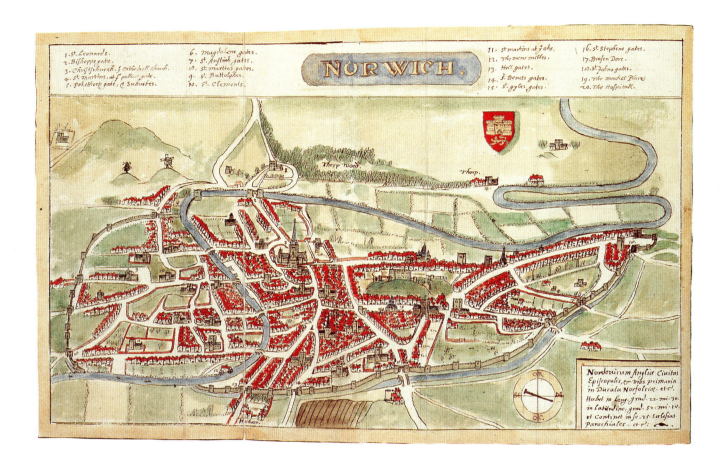

where, producing printed maps that were made the same way with pictures superimposed on a scale-plan. Most were much smaller in size, but two showing the university towns, each known from a single copy, measured nearly 3 feet by 4 feet: Oxford, printed from eight plates, drawn by Agas in 1578 and engraved and published ten years later, and Cambridge, printed from nine plates in 1592, from drawings probably by John Hamond but also attributed to Agas. Another plan of Cambridge, by an unknown map-maker but engraved by Richard Lyne, had been published in 1574; though of more modest size it illustrates, significantly, a book by John Caius arguing that Cambridge was the older university – again, local pride was to the fore. A plan of Exeter engraved by Remigius Hogenberg, brother of Frans, was published in 1587 at the expense of John Hooker, chamberlain of the city, and by 1600 Braun and Hogenberg's collection, besides London, included plans of Bristol, Canterbury, Chester and Norwich, while some other places in England were represented by simple views.

Braun and Hogenberg probably drew on existing manuscript plans. A map drawn to enhance a town's prestige did not have to be printed, and this may be

fig. 7

52 Norwich, by William Smith, 1588
A map with superimposed pictures, one of several illustrating Smith's 'Particuler Description of England', unpublished until 1879. South-east is at the top of the map, but the cathedral has been turned round to be viewed as though from the south.
British Library, Sloane MS. 2596, f.61

75

Maps and Towns

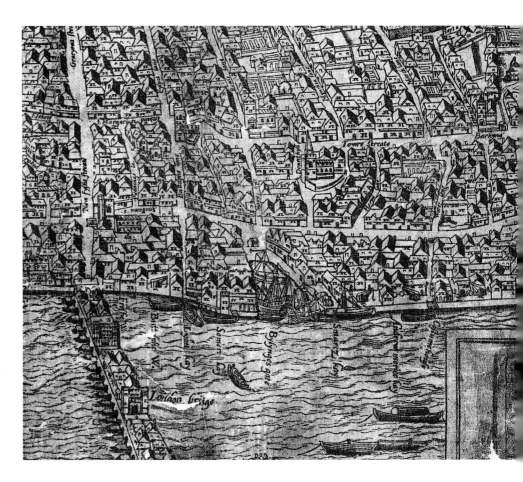

53 London, late sixteenth century
Part of a large map, 2 by 6 feet, printed from eight woodblocks. The only surviving copies date from the seventeenth century and show signs of revision, but it was probably originally based on the engraved map of 1553–9 (no.51). Looking northwards from the Thames, the Tower of London is on the right, London Bridge – with its many piers and built up with houses – on the left.
Public Record Office, MPEE 25

the origin of some of the plans or bird's-eye views of towns that came into the possession of the government. One, of Newcastle seen from south of the Tyne, is perhaps connected with the defence works at Tynemouth in 1545, but others may well have been copied, to be used for reference, from originals that had been made for town authorities, perhaps to adorn a guildhall or other civic rooms. This is no more than a guess, but it is a possible explanation of, for instance, a plan of Shrewsbury that Burghley bound with his proof set of Saxton's county maps, or the plan of Great Yarmouth, painted probably in the 1580s or 1590s, that survives among maps from official records.

fig. 47
fig. 48
fig. 9

By the end of the sixteenth century a genre of town plans had been established in England, and with it a clear distinction between the straightforward view and the map, which might have all features shown pictorially but which was based firmly on a plan that was at least notionally drawn to scale. It was now that town plans began to be seen as a natural adjunct to county maps. There are no town plans in Saxton's atlas – indeed, we know of no town plan

Maps and Towns

ever drawn by Saxton. But John Norden included plans of Higham Ferrers and, probably, Peterborough in the copy of his description of Northamptonshire that he presented to Burghley, and although there are no town plans on the abortive series of county maps that William Smith produced in 1602–3 he was experienced in this branch of cartography. His account of Cheshire, written in 1585 but not published, includes a map of the county, views of Chester, Halton and Beeston, and a plan of Chester, while another manuscript, which he wrote in 1588, is entitled 'The Particuler Description of England With the Portratures of Certaine of the Cheiffest Citties and Townes'. It includes detailed maps of Canterbury, Bath, Norwich, Cambridge and Bristol, and views of some other towns, but aspiration outran achievement, for both are far outnumbered by the spaces left blank for maps or views that were never drawn. However, both Norden and Smith were paving the way for John Speed, whose county maps, engraved in the early years of the seventeenth century, include as insets a magnificent collection of over seventy plans of towns in Britain and Ireland.

fig. 52

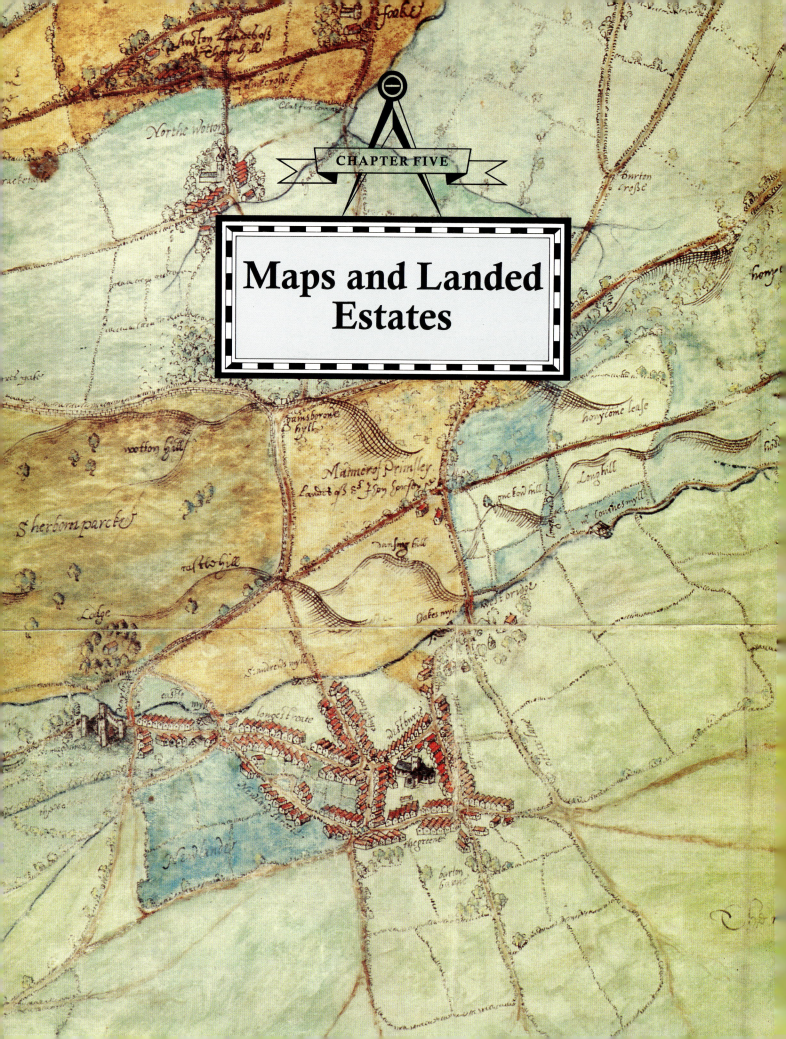

CHAPTER FIVE

Maps and Landed Estates

Maps and Landed Estates

MOST of the 20-odd local maps from fifteenth-century England are concerned with the internal or external problems of landed estates, showing how contiguous parcels of land adjoined, setting out disputed rights or boundaries, and so on. When the production of maps escalated after 1500 it was particularly maps of this sort that were drawn – the Elford map of 1508 is an example, and so is the plan of the roads and fields near Andwell some years later – and they continued to be produced at, probably, ever increasing pace. One, which cannot be earlier than 1537, shows the rights of five adjacent proprietors in marshlands on the north Kent coast at Cliffe, north of Rochester. It is a mixture of diagram and picture – the tracts of marsh are shown as a row of rectangles with the manor house of each at the top and the River Thames at the foot with boats and fishes. Another, of lands around Cotehele House in Cornwall, may have been drawn in the 1550s to show how access to two mills and other paths might be maintained when a deer-park was created. It is a simple sketch-map, with rough drawings of gates, trees, Calstock church, Cotehele House itself and other buildings.

It seems that by the mid-sixteenth century it was quite normal for landowners, or anyone concerned in managing landed property, to draw a map if ever occasion arose when it would be useful – something all but unheard of a hundred years earlier. These maps vary enormously in style and appearance, in the care taken in drawing them, in the extent and character of the lands they show. Two features, however, are common to them all. One is that they are all diagrams or picture-maps or, like the Andwell plan, simply pictures – there is no consistent scale, and features above ground-level are shown not in outline plan but pictorially, like the houses on the plan for Cliffe or the various features drawn on the Cotehele map. The other is that each map was drawn *ad hoc*, to serve a particular purpose on a particular occasion – it does not seem to have occurred to any landowner to have a map of his estate for general reference, a map that he would keep permanently so that he could use it to look up any point as it arose.

The records that estate owners did keep for permanent reference were not maps but surveys, written descriptions of their properties, mostly in set form, with measurements of each piece of land and precise definitions of the rents due to them and of the rights they could enforce over land and tenants. These surveys had a long history. They survive from the tenth century onwards and in the thirteenth century they begin to give the measured area of each piece of arable. Some medieval surveys are amazingly full and detailed – they might define the location and size of every one of hundreds, even thousands, of pieces of land scattered throughout a manor's open fields. From the thirteenth century onwards some men seem to have developed particular skills in measuring lands, and to have been specially brought in when surveys were drawn up, but it

Maps and Landed Estates

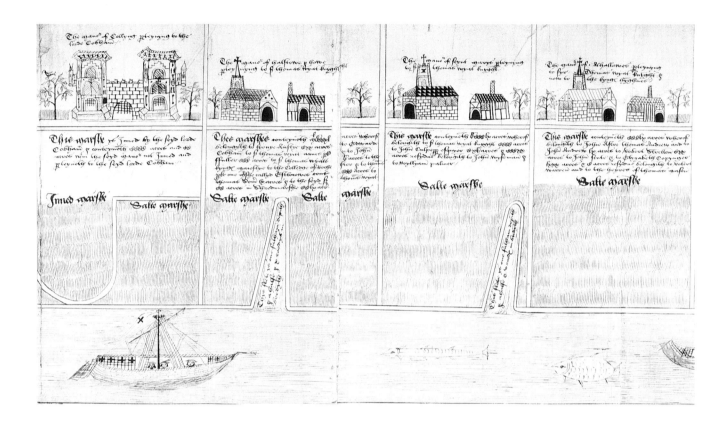

54 Villages and marshland on the north Kent coast, mid-sixteenth century
Part of a map showing Cliffe and the succession of places eastward – Cooling, Halstow, St Marys, Allhallows. The River Thames, lying to the north and separated from the settlements by their tracts of marsh, is at the bottom of the map. As Cliffe, the westernmost village, is at the left end, this is another example of a map drawn in mirror image, like no.1.
British Library, Harley MS. 590, f.1

was only in the sixteenth century that fully qualified surveyors emerged, men who made it their business to measure and evaluate landed property, describing the results in written surveys. The medieval surveyor had the help of specimen surveys and of tables to work out the areas of lands, and the first printed treatises on surveying appeared early in the sixteenth century – the *Boke of Surveyeng* by John Fitzherbert in 1523 and the *Maner of Measurynge* by Richard Benese in 1537.

In running his estate the mid-sixteenth-century landowner might thus draw on two quite unconnected activities. One was the long tradition of having written surveys of landed property made by surveyors, who were becoming increasingly professional in their approach and techniques. The other was a growing tendency, whenever circumstances demanded, to draw maps which ranged from simply sketched diagrams to elaborate, carefully worked picture-maps. Then, in the 1570s and 1580s, these two activities began to coalesce. Maps of estates began to be drawn by surveyors either to supplement or to replace their written surveys, the start of a process that eventually made surveying all but synonymous with map-making. The earliest known seem to be Ralph Agas's map of West Lexham in Norfolk in 1575 and two sets of maps by Israel

Maps and Landed Estates

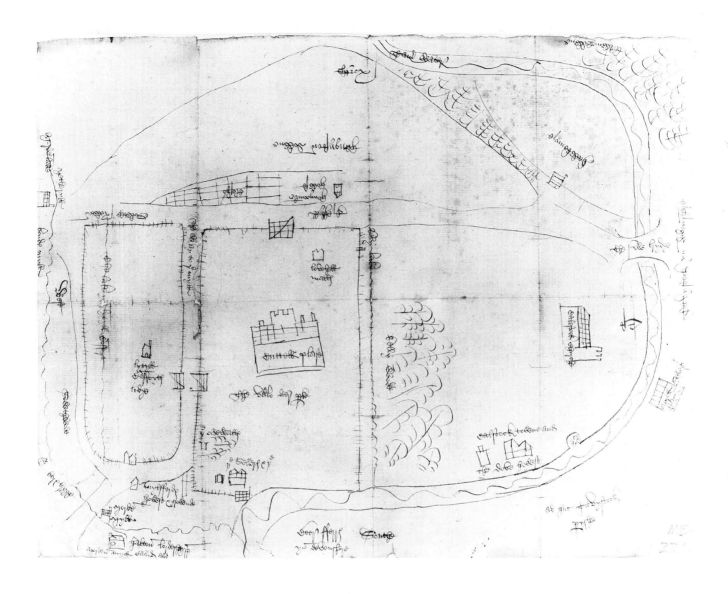

Ames of properties in Essex: the manor of the dean and chapter of St Paul's Cathedral, London, at Belchamp St Paul in 1576 and several manors and other lands of Edmund Tirrell in 1579. It is likely that earlier examples will come to light of estate maps drawn by surveyors for general reference; it is unlikely that they will be much earlier than these. Certainly it was only in the 1580s that estate maps became established as a regular genre of English cartography, and by the end of the century they were still far from being a normal tool of estate management.

In 1596 Agas told how, some 20 years earlier, he had been employed to trace certain bounds and boundary marks of a property, and having done this 'I

fig. 58
fig. 59

55 Cotehele, Cornwall, mid-sixteenth century

A sketch-map perhaps drawn on the creation of an enclosed deer-park to show the access to the three mills along the left edge. Cotehele House ('Cuttell plase') is in the centre, Calstock and its church to the right, beside the River Tamar. Before the 1570s, maps of landed properties seem always to have been drawn to meet a particular purpose, not for general reference.

Cornwall Record Office, DD ME2369

Maps and Landed Estates

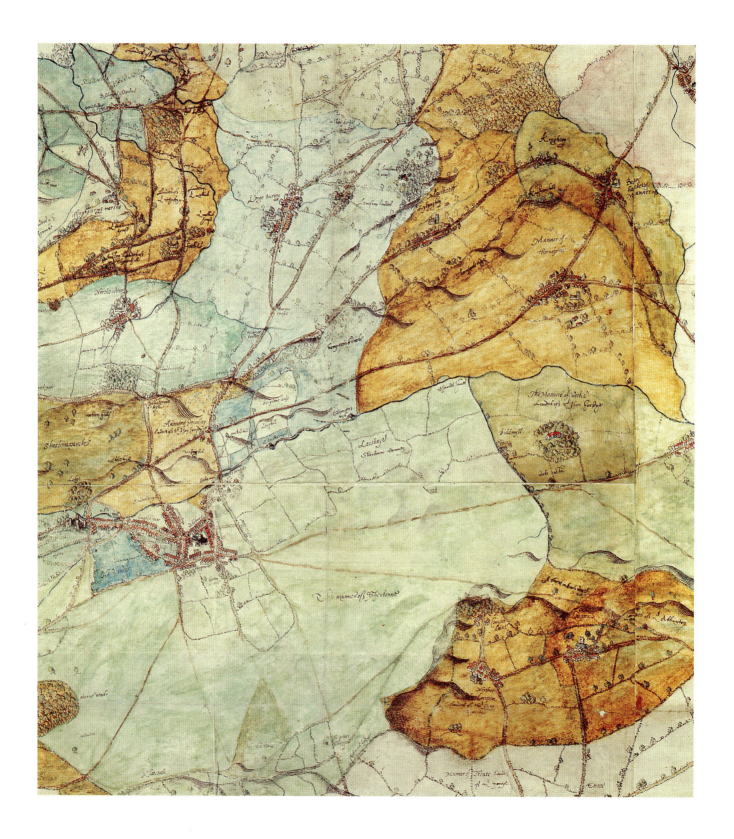

considered of what force a bounder by plat might be in time to come' – drawing a map, in other words, had been no part of this particular surveying job, but it had occurred to Agas how useful it might have been to draw one for future reference. We know nothing of how the first estate maps came to be drawn, but it seems unlikely that it was because surveyors only now had the idea of drawing them. They must have been long aware of the possibility. For one thing, drawing maps may well have been a normal part of the surveyor's work, not as the finished product but as a means of assembling and setting out the information that was to be put in the written survey. Even from the fifteenth century we have one or two sketch-maps that are best explained this way, and, to take a much later instance, some sketch-maps from Dedham in Essex, dated August 1573, give the area and occupier's name for each of the many plots of land they show, as well as page references to unidentified records, and are probably working notes for a written survey of the manor by Robert Mawe, which the next year was described as nearly finished. We have seen that every map had two parents, the map-maker who drew it and his client or patron who commissioned it, and it was probably the landowners who effectively brought about the introduction of estate maps, allowing themselves to be persuaded by the surveyors that maps of their properties were worth having – and worth paying for.

fig. 57

The treatises on surveying that succeeded Fitzherbert's and Benese's in the course of the sixteenth century confirm this chronology for the estate map and show that it was not the immediate result of new techniques or technology. *A Boke named Tectonicon* by Leonard Digges, published in 1556, gives directions for measuring surfaces – wood, stone and glass as well as land – and describes an 'Instrument Geometrical' or 'Profitable Staffe' that could be used, but he does not mention the possibility of drawing maps. Another book by Digges, which his son Thomas – the future engineer of Dover Harbour – completed in 1571, *A Geometrical Practise, named Pantometria*, envisages using quadrant, theodolite and other instruments in drawing scale-maps, but these maps were for military operations, not estate surveying, as both the text and the illustrations make clear. Similarly William Bourne in 1578, in his *Booke called the Treasure for Traveilers*, tells how to draw regional maps, but says nothing of map-making in the section on measuring lands, while Valentine Leigh, whose *Moste Profitable and Commendable Science, of Surveying of Landes, Tenementes and Hereditamentes* appeared the previous year, is concerned mostly with the form of the written survey and other records, and though he remarks – reasonably enough – that 'the Surveiour should have some skill in measuryng of Lande' and tells him how to do it, he says nothing of drawing maps. The first treatise to suggest that a surveyor might draw maps of an estate was published in 1582: *A Discoverie of Sundrie Errours and Faults daily committed by Landemeaters, Ignorant of Arithmetike and Geometrie*, by Edward Worsop. The two instruments, with various refine-

56 Sherborne, Dorset, and surrounding area, 1569–78
This is only a part of the map, which covers an area some 9 by 10 miles. Its precise date and purpose are uncertain, but it was drawn for the bishop of Salisbury, lord of Sherborne itself and of nearby manors. Overall colouring distinguishes each manor – both the bishop's and other lords' – and the landscape is pictured in fine detail. South is at the top and Sherborne, with its castle and abbey church, is on the left.
British Library, Additional MS. 52522

ments, that surveyors used in the late sixteenth century to measure land – and to draw maps – were the plane-table, on which the directions of features were drawn by simple sighting in the field, and the theodolite, from which the measurements of angles could be read for subsequent calculations in the study. It is likely that both were regularly used by surveyors for land measurement long before they started to draw estate maps.

From the very start estate maps were drawn to scale. Given that they were drawn by surveyors, this is not surprising: having measured the lands with plane-table or theodolite, they will have had at hand all the information they needed to draw a scale-map. What is more surprising is that landowners were prepared to accept these scale-maps and did not demand picture-maps instead. There were of course clear advantages for the landowner in having his estate map drawn to scale. The picture-map would tell him no more than he probably knew already, though to the absentee landlord it might of course be at least a useful *aide-mémoire*. The scale-map, however, would show him at once, far better than the written survey, how his various pieces of land compared in size and position, and the precise effect of dividing or combining them for his own cultivation or for letting or for sale. But this would require a familiarity with the idea of scale that could not be taken for granted in the late sixteenth century. We have seen (p.65) how George Owen thought – rightly or wrongly – that even the queen's counsellors had so little understanding of the different scales in Saxton's atlas that they misjudged the relative sizes of the counties of southwest Wales. Edward Worsop's treatise of 1582 takes the well-established form of a conversation, in this case between a surveyor and four other men. One of them asks what the word *scale* means, 'which I see in so many places of your booke'. After a long explanation, illustrated with pictures of scale-bars and dividers ('compasses') the questioner is finally satisfied: 'I nowe perceive what is meant by this worde scale. I remember I have seene the like lines, and compasses set in mappes, but I never understood what they meant till nowe'.

The landowners who had estate maps drawn must somehow have achieved this understanding of scale if they were to use the maps at all intelligently, and the simplest explanation is that they had been introduced to the idea by Saxton's county maps. The coincidence of date is striking – the earliest estate maps were drawn while Saxton's maps were being produced, between 1574 and 1578 – and the owners of estates were just the sort of men whom we might suppose would have bought the county maps, being much involved in county society, county administration and county justice. Having seen the value of such a map – how, drawn to scale, it could be used to measure any distance within the county – they could easily be persuaded of the value of a map of their own estates. Nor was it a matter simply of practical use. Just as the county map, with its embellishments of allegorical figures, fruits and flowers, symbolised the

Maps and Landed Estates

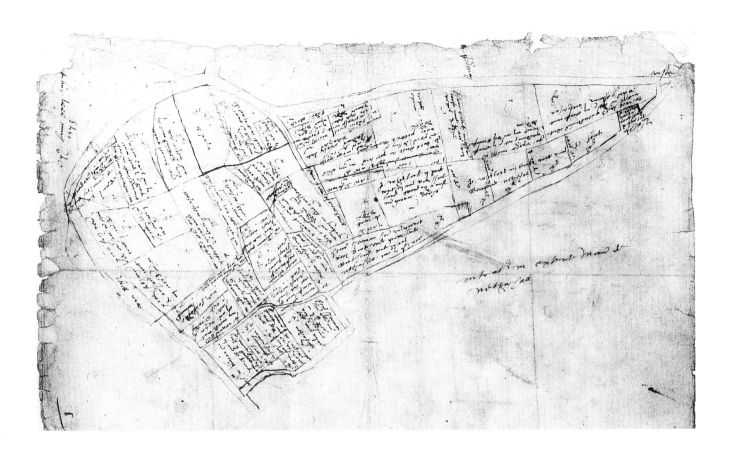

county, representing it in miniature, so the estate map, with its tiny pictures of manor house and tenants' houses, of mills and barns, trees, gates and fences, was a statement of ownership, a symbol of possession such as no written survey could equal.

In other ways too, besides being drawn to scale, the estate map emerged fully fledged from the start and underwent very little development or improvement over the years. Some of the earliest estate maps, like many later ones, accompanied and illustrated written surveys. The maps of Belchamp St Paul in 1576 come at the end of a manuscript volume of some 90 leaves containing Israel Ames's full written survey of the manor; there are seven plans of individual closes, then a much larger map of further demesne lands and woods, then a still larger map of the whole manor. The same cartographer's eight maps of Edmund Tirrell's lands in 1579 also accompany a written description, but are interspersed in a much smaller book. On the other hand, the earliest estate maps include some that completely replaced the written survey, setting out all the information either on the map itself or in panels around the edge. Ralph Agas's map of West Lexham in 1575 was of this kind – it includes a note of

57 Fields at Dedham, Essex, 1573
One of a group of sketch-maps probably used to compile a written survey. On each piece of land are noted the tenant's name, its area and the manor to which it belongs, Overhall or Netherhall. Surveyors of landed property seem to have used sketch-maps as a way of noting information long before estate maps began to appear alongside written surveys as the end-product of their work.
Public Record Office, MPC 77, f.4

Maps and Landed Estates

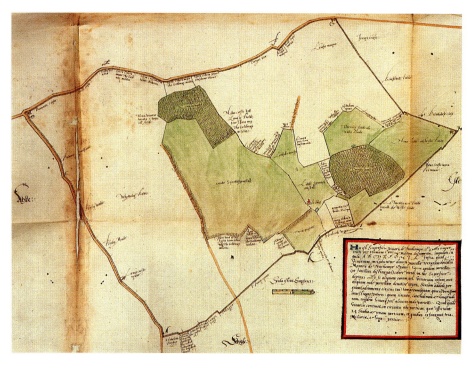

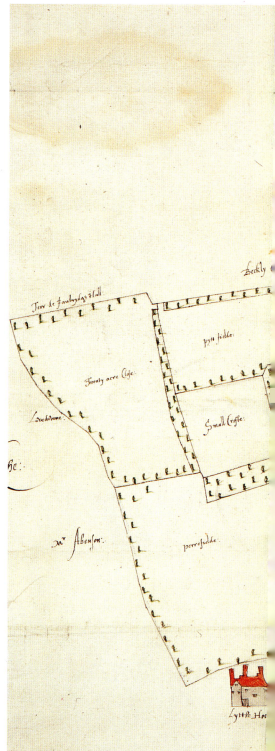

58 Belchamp St Paul, Essex, by Israel Ames, 1576

The written survey of this manor of St Paul's Cathedral by Israel Ames was one of the earliest to be illustrated with maps. This map shows a park, its bounds marked by the letters A (at the bottom) clockwise to K. The Latin note (bottom right) explains how, using the scale-bar and dividers (*circenus*), areas and lengths can be worked out from the map.

Guildhall Library, London, MS. 25517/1, f.100

59 Plumberow, Essex, by Israel Ames, 1579

One of eight maps of properties of the late Edmund Tirrell. They accompany a written survey made so that the estate could be divided between his heirs. The map gives a clear picture of the landscape of south-east Essex, with scattered farms surrounded by enclosed fields and patches of woodland.

British Library, Harley MS. 6697, f.19

Maps and Landed Estates

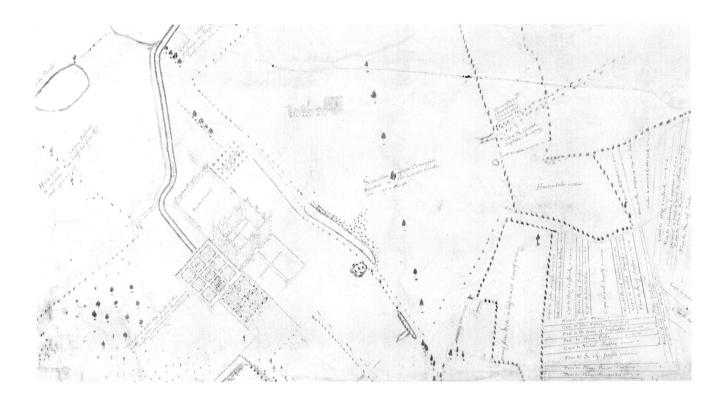

60, 61 Toddington, Bedfordshire, by Ralph Agas, 1581
Two tiny portions of a masterpiece of English estate-mapping, drawn when the genre was in its infancy. Agas's map of Lord Cheney's estate, on the scale of about 40 inches to the mile, is on twenty sheets of parchment and measures more than 11 by 8 feet.

60 On the left is Lord Cheney's house with its gardens and grounds; in front of it are enclosed meadows and open-field arable, divided into the strips of the individual tenants.
British Library, Additional MS. 38065 H

61 Agas's map of Lord Cheney's estate at Toddington includes what is probably the best picture we have of a small Elizabethan country market-town – telling us too who owned every house and the area of every patch of ground. The hill on the right is where the Norman castle once stood.
British Library, Additional MS. 38065 N

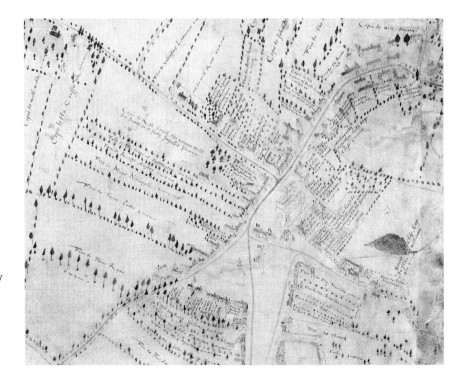

manorial customs and a list of tenants with their rents – and so too was the map of Lord Cheney's manor of Toddington in Bedfordshire that Agas drew in 1581.

This is not only an outstanding cartographic monument of the sixteenth century – it is also one of the most remarkable of all English estate maps, and it is interesting that so ambitious and so sophisticated a map was produced while the genre was still in its infancy. It is a large map – some 11½ by 8½ feet – drawn on 20 sheets of parchment and shows the whole manor at a scale of about 40 inches to the mile. Every piece of land is distinguished, including innumerable strips in the open fields, and on each are noted its area, the name of the occupier and the form of tenure – freehold, copyhold, glebe and so on. The village with its church and market cross, Lord Cheney's house with its gardens and ornamental grounds, have all been drawn from life and imposed on the map in bird's-eye view, and many other features are shown in more or less conventionalised pictures. In one corner are summarised lists of the lord's own lands and of the holding of each tenant. The whole can be seen – and will have been seen by Lord Cheney – as a written survey set out in a graphic form that not only told him more than any description solely in writing could do, but also set out the size and value of his estate for all to behold.

How the landowner might use his estate maps is suggested by an interesting set of maps of the manors of Sir Christopher Hatton in Northamptonshire; they accompany written surveys and were drawn at various dates between 1580 and 1587 by Ralph Treswell. For two of the manors, Holdenby and Kirby, there are two maps, the second showing the effects of enclosure, imparking and exchanges of land that had been made since the first was drawn. These second maps were produced – in 1586 and 1587 – probably because Treswell was still at work on other parts of the estate and it seemed sensible for him to re-map these altered areas so that the whole series of maps was up-to-date; they do not seem to have been drawn in advance to show the effects of the projected changes. But we can see at once – as Hatton must have seen – how easily he could plan these changes on the estate simply by referring to the maps. This again was something no written survey could do.

It was not, however, all landowners who saw these advantages – or who, having seen them, thought them worth the considerable expense of having maps drawn of their properties. The written survey still had a long history before it and, indeed, there were professional surveyors in the late sixteenth century who offered their clients no more than this. The map-makers among the surveyors appear in their writings as a band of enthusiasts, anxious to open the eyes not only of the landowners but of their own colleagues as well to the advantages of estate maps. In an unpublished note Agas wrote that 'No man may arrogate to himselfe the name and title of a perfect and absolute Surveior of

Maps and Landed Estates

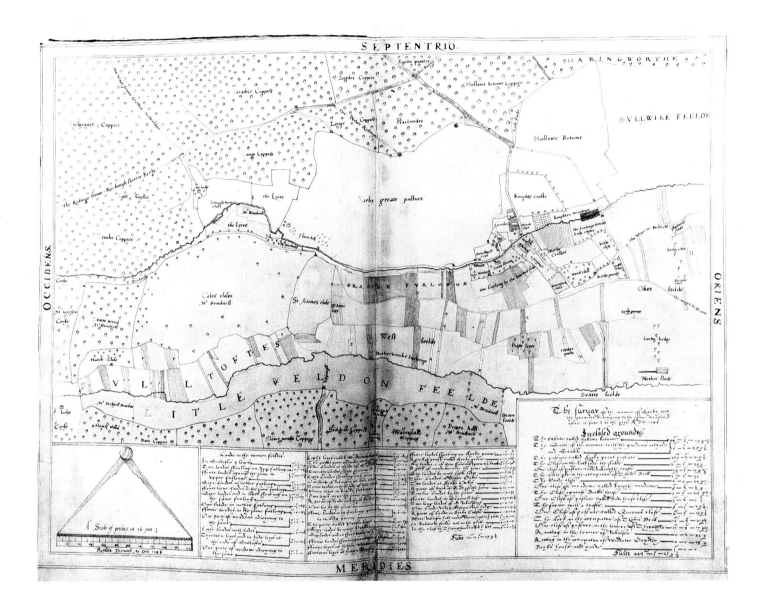

62, 63 Kirby, Northamptonshire, by Ralph Treswell, 1585 and 1586

Between 1580 and 1587 Sir Christopher Hatton had a series of maps of his estates drawn by Ralph Treswell. These two are both of the manor of Kirby, a now deserted site near Corby.

62 Drawn in 1585, this map shows much of the land between the two brooks divided into open-field strips.
Northamptonshire Record Office, Finch Hatton MS. 272, f.5

63 On the revised map of 1586 the strips have disappeared and the notes below recount the exchanges and enclosures by which this was effected.
Northamptonshire Record Office, Finch Hatton MS. 272, f.6

Maps and Landed Estates

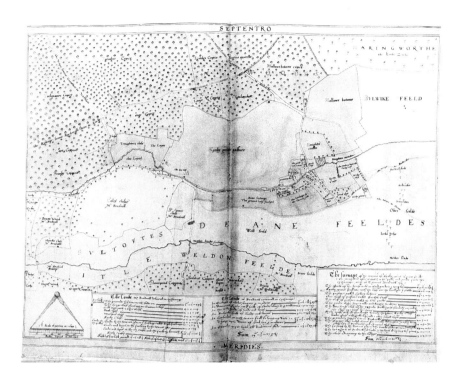

Castles, Manners, Lands, and Tenements, unlesse he be able in true forme, measure, quantitie, and proportion, to plat the same in their particulars ad *infinitum*' – that is, unless he could draw detailed scale-maps of these properties. John Norden, in *The Surveyors Dialogue* published in 1607, puts the point of view of the farmer who sees no need for his landlord to have the property mapped: 'is not the Field it selfe a goodly Map for the Lord to looke upon, better than a painted paper? And what is he the better to see it laid out in colours?'. The answer was given not only by Norden but – more forcefully – by Agas in his treatise of 1596, *A Preparative to Plotting of Landes and Tenements for Surveigh*: the bounds of each piece of land could be shown in greater detail on a map than in a book, if land is to be divided the scale-map will show 'where may be the rediest cut, and with what charge accomplished', the map or 'surveigh by plat' avoids any risk of losing lands that have acquired new names in the course of time – and so on.

One of the surveyors who did draw maps in the 1580s and 1590s was Christopher Saxton, and his career is instructive. We know of no map that he drew, nor any written survey that he compiled, earlier than his maps of the counties – nor, indeed, for some years after their publication, and the publication too, in 1583, of his large wall-map of England and Wales. In 1583 and 1587 there are references to his making surveys – including a map – of estates in Yorkshire and Kent, and from then until his death in 1610 or 1611 he seems to have worked

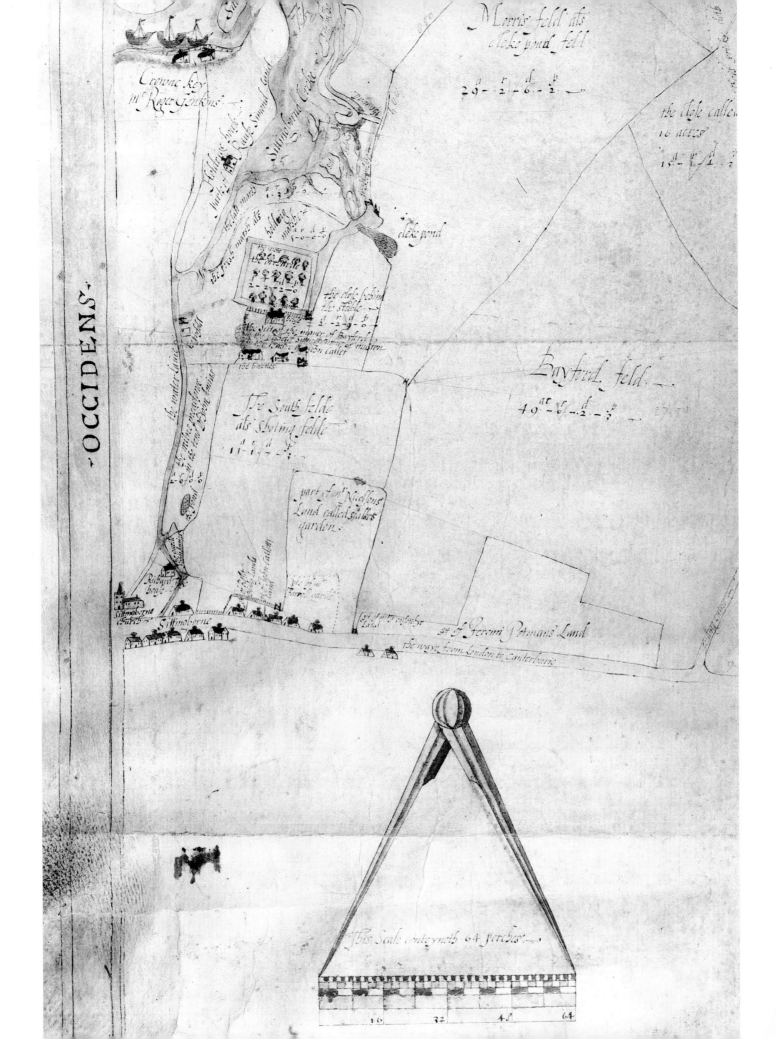

consistently as an estate surveyor producing both maps and written surveys. His earliest surviving estate maps are from Kent in 1590, but latterly he worked mostly in his native Yorkshire. He may or may not have learned estate surveying as well as cartography from John Rudd, and may or may not have worked as a surveyor, producing written surveys, before Seckford commissioned him to map the counties. What his career shows, though, is how easily the distinct crafts of surveyor and map-maker combined and fused, even when the map-making was of not even the largest landed estate but of the entire realm.

All the same, even so keen an advocate of estate mapping as Norden did not see surveyor and map-maker as synonymous. In *The Surveyors Dialogue* he denies that 'plotting is the chiefe part of a Surveyors skill' – it is merely one skill he should have along with others 'more necessary'. These included mensuration, Latin and the ability to read ancient records, knowledge of law and of property valuation. Even when the survey resulted in a map it was made not only with measuring rod or chain, plane-table or theodolite. According to the title-panel on the map, Agas made the 1581 Toddington survey 'aswell by the vue off divers good & auncient Evidence as also by the Oath and Information' of 24 local tenants, all named, taken in a court of survey, a special meeting of the manorial court. This was normal practice, which long persisted – in the nineteenth century, books on holding manorial courts still describe what to do at a court of survey. The lands might be plotted and measured in the field, but to discover who held them and on what terms the surveyor had to turn to the manorial records and the tenants themselves. The surveyor may have become a map-maker, but he did not stop doing what surveyors had done before. Even the map that entirely replaced the written survey is best understood as a written survey of a peculiar sort – its title and the information it gives all derive from the traditional manorial survey. This is a boon for the historian of cartography, for just as the heading of a written survey normally gives the date and often names the surveyor so the estate map usually bears both a date and the name of the map-maker – in contrast to the undated and anonymous early maps of fortifications and coastal defences.

By the end of the sixteenth century estate maps were well established, whether supplementing written surveys or replacing them altogether. But for a long time to come it was only certain landowners who had them made – others did not. No one group of landowners emerges as especially favouring maps, and the decision whether or not to have them may have rested on the funds available, on individual opinions of their value or on simple chance. Thus among Oxford colleges, All Souls and New College commissioned estate maps in the 1580s and 1590s, whereas Corpus Christi employed Agas in 1583–5 only to make written surveys. The fusion of surveyor and map-maker had made a good start, but still had far to go before it was complete.

64 Sittingbourne, Kent, by Christopher Saxton, 1590
Part of one of the earliest estate maps by Saxton, showing the two manors in Sittingbourne owned by Sir William Garrard. It is not in Saxton's own hand and was perhaps drawn by an assistant. Part of the town, with the church, appears on the left. The areas of the fields are given in acres, roods, dayworks and perches – the daywork of 4 perches was a local measure in Kent and Essex. Here and in other maps dividers are drawn across the scale-bar by way of instructions, showing the user how to measure distances on the map.
British Library, Additional MS. 50189

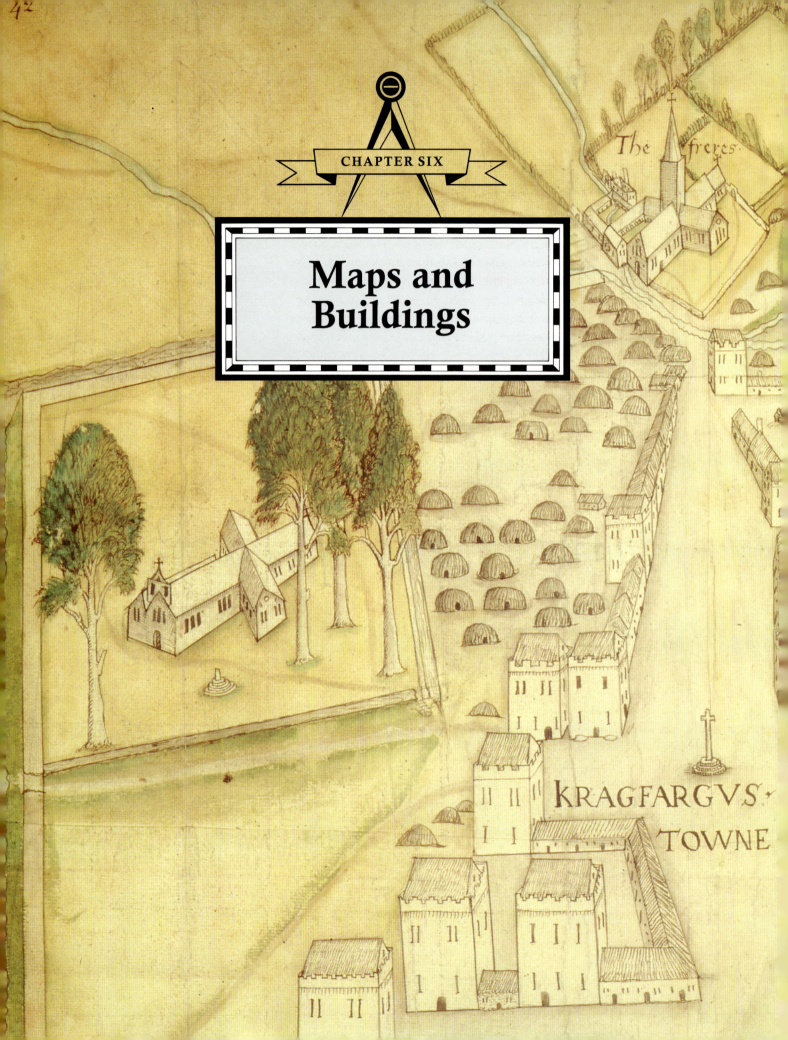

CHAPTER SIX

Maps and Buildings

Maps and Buildings

fig. 8
fig. 68
fig. 60

BUILDINGS of every kind appear on Tudor maps, from the humble field-barn near Andwell and the bee-hive huts of Carrickfergus to the splendid home of Lord Cheney at Toddington. And they appear in many forms – simple ground-plans, symbolic outlines of houses or churches, bird's-eye views and other pictures that might be conventionally drawn or might be realistic and detailed. But besides this there was a tradition of plans of individual buildings, its development quite distinct from that of other plans and maps, but closely related. Like every branch of sixteenth-century cartography they pose questions that have still to be answered, and the crucial question here is the purpose of these plans – what part, if any, did they play in the actual construction of buildings? Were they technical drawings, by which master masons – the architects – communicated their intentions to their underlings? Or did plans only gradually come to play a part in the builder's craft, a craft that had hitherto managed without them? Does the appearance of scale-plans of projected buildings in the sixteenth century reflect – or did it initiate – a change in the organisation and techniques of construction? Eventually we may know the answers to these questions – at present we do not.

The earliest known reference in England to what was probably a building plan is in a contract of 1380 for putting up buildings in Hertford Castle, 'as set out in a design (*patron*) drawn in an indenture' – an indenture of which client and builder would each keep one part. Other later medieval contracts refer to plans, drawn by the builder for the client, to which the finished building was to conform, and a book of Latin and English sentences by William Horman, vice-provost of Eton College, which was published in 1519, says of the builder that 'He is not worthy to be called maister of the crafte, that is not cunnyng in drawynge and purturynge (*i.e. portraying*)' – foreshadowing what Agas was to say of the surveyor some 70 years later. Although patterns of mouldings survive from the fourteenth century onwards, we have only a solitary building plan earlier than 1500 – indeed, it is only a fragmentary part of a plan, drawn about 1390 and preserved by being drawn on parchment that was later re-used in binding. It shows one corner of the inner court of Winchester College, with a doorway (in elevation) and a flight of stairs (in plan) and is not drawn to scale but has a note of the size of one of the rooms, the new kitchen, 30 feet long and 20 feet broad. From the 1470s we have a group of four outline plans of pieces of land in London and at Deptford and Lambeth in Surrey, drawn in a register of properties belonging to London Bridge. They are not building plans – a house built on one of the plots is drawn in elevation – but they use the same technique as the Winchester College plan in being drawn roughly the right shape but not to scale and with the exact length of every line on the ground written beside it on the map. Three of the plans call themselves *patron* ('This is a patron...', etc.), the fourth calls itself a *platte*.

Maps and Buildings

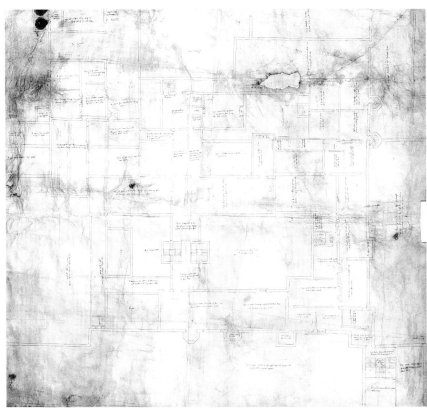

65 Bishop Fox's chantry chapel, Winchester Cathedral, *c.*1526
An elevation, drawn to scale, of one of the chapel's four bays. Differing in detail – and in its proportions – from the finished work, it is probably a preliminary design. There is some reason to suppose that isometric elevations came into use among master masons and their patrons earlier than ground-plans drawn to scale.
Royal Institute of British Architects, Smithson Collection, Box 3, AE 7/15

66 Exchequer at Calais, 1534
A carefully drawn plan to show how the building was arranged for the meeting there of the kings of England and France. All 60 rooms, and the courtyards between them, are identified and many dimensions are given – 'The frenche kyngis kychyn 30 foote Longe and 22 fote Brod' – but these show that it is not drawn to consistent scale. Straight and spiral staircases are carefully distinguished.
British Library, Cotton MS. Augustus I, Suppl. 7

Probably this was the normal form of the building plans from medieval England – drawn to shape but not to scale, and with all relevant dimensions written on the plan. Certainly building plans took this form in the early sixteenth century and much later. When in 1538 the Lord Mayor of London wrote to Thomas Cromwell, mentioning 'a platte, that was drawn howte (*i.e. out*) for to make a goodly Burse in Lombert Strette for Marchaunts to repayer unto', this may well have been a ground-plan of this sort, for they were used for important and elaborate buildings. One outstanding example shows how the Exchequer building at Calais was adapted to accommodate the kings of England and France when they met there in 1534. Measuring more than 4 feet square, it shows some 60 rooms grouped around a series of courtyards, with adjacent streets and gardens. It is drawn entirely in plan – walls are shown by double lines, doorways are marked with their jambs, straight stairs appear as series of parallel lines, spiral stairs as radii from a central hub. There are many measurements on the map, but it is not drawn to scale. Clearly, however, a plan as elaborate as this would fail to show features in proper relationship if it diverged far from a more or less consistent scale, and we find that in the course of time these ground-plans might record the outline shape so accurately that they approached closely to scale-plans. This can be seen in four plans drawn by Henry Hawthorne in 1576 for a new gallery at Windsor Castle. They are finely drawn in ink, with some shading in wash, and with partial consistency of scale – on one a room 20 feet wide is drawn exactly twice the width of one of 10 feet, but a walk shown as 16 and 17 feet wide is wider than the widths of rooms 10 feet and 6 or 7 feet added together. None of the four has a scale-bar or note of scale.

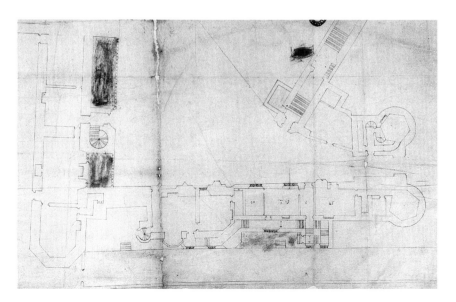

67 Proposed new rooms at Windsor Castle, 1576
Part of one of the plans drawn by Henry Hawthorne as a basis for estimating building costs. It is drawn with great care, and measurements have been written in, but it is not to scale – the two passages marked '4–6' are exactly the same width, but the room marked '21' is disproportionately wider than the room marked '19'.
Public Record Office, MPF 150

Maps and Buildings

Some building plans had been specifically drawn to scale long before this. Earlier still, however, we have a very few isometric elevations of buildings – drawn to consistent scale, with no more than a touch of perspective to mark protruding masonry. One, especially striking in its elaborate architectural detail, shows a bay of Bishop Fox's chantry in Winchester Cathedral, built some time before 1528. Even if drawn after the building was completed, the picture is unlikely to be much later than this. It is interesting that in the magnificent late-medieval collections of architectural drawings from south Germany and Austria the earliest elevations drawn to scale similarly predate the earliest scale-plans – they date from the late fourteenth century, whereas the plans are some decades later. It may be – though this needs to be tested by research in other areas – that this is a consistent pattern and that isometric drawing is more easily comprehended in elevation than in plan.

fig. 65

The way scale-plans of buildings were introduced in England is of especial interest because, occurring in two stages, it reflects the way other scale-maps were introduced, first to a limited range of knowledgeable persons, then to the wider public. The first stage comprised plans of civil buildings drawn by military engineers. There is nothing surprising in this. The distinction we would make between the arts of war and the arts of peace scarcely applied in the sixteenth century. A ruler would expect a fortress or the defences of a city to be imposing, monumental and aesthetically pleasing – the same qualities that he would look for in a palace or any other considerable building. Fortification was simply architecture meeting a particular set of requirements, the requirements of defence, just as in the palace it would meet the requirements of domestic comfort and household efficiency. We have seen that our earliest scale-plans of fortifications were drawn in 1540, by or for Richard Lee and John Rogers, and that they probably learned the technique from one or other of the Italian engineers who were also in Henry VIII's service. But although Lee and Rogers were employed to plan and build defences they were simply master masons – architects – who will not have seen fortifications as different in kind from any other building. Nor did their employer, and when in 1541 Rogers was at Hull, planning the town's eastern defences, he also had to plan alterations to the manor house that the king had acquired there two years before. These alterations included building a bastion and gunports in the perimeter wall of the property, so that it would 'serve as a Citadell', as Rogers showed in a perspective view – but they also included internal changes to fit the house for occupation by the king's and the queen's households. These appear in three plans, one of the house unaltered, the others showing the changes on the ground floor and the first floor. Two have dimensions written in every room, but all three are drawn to consistent scale. They are our earliest scale building plans from England.

fig. 69

fig. 70

68 Carrickfergus, County Antrim, *c.*1560
Many kinds of buildings appear on Tudor maps, but this presents an unusual range – the church, friary and castle, the stone towers with rows of houses between them and the cabins and beehive huts behind. The jetty and boats in the foreground are drawn with minute attention to detail.
British Library, Cotton MS. Augustus I.ii.42

We have to wait another 30 years before we find scale building plans wholly

98

Maps and Buildings

69, 70 View and plan of the king's manor house at Hull, 1542

John Rogers, king's engineer, in Hull for work on the town's defences, had also to make alterations to the manor house that the king had recently acquired there, adapting it both for defence and as a royal residence.

69 Isometric view, showing gunports at the angles of the perimeter wall, and a proposed bastion in front.
British Library, Cotton MS. Augustus I.ii.13

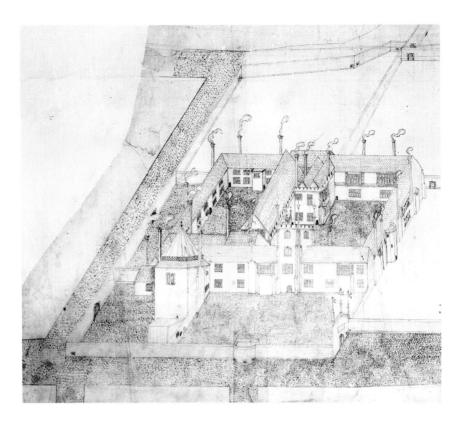

70 Plan of the proposed alterations on the first floor. Though not specifically stated, it is drawn on a scale of about 1 inch to 16 feet, like two other plans of the alterations. The three are the earliest known plans of civil buildings in England that are drawn to scale.
British Library, Cotton MS. Augustus I.i.84

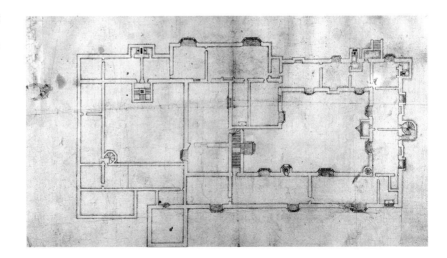

Maps and Buildings

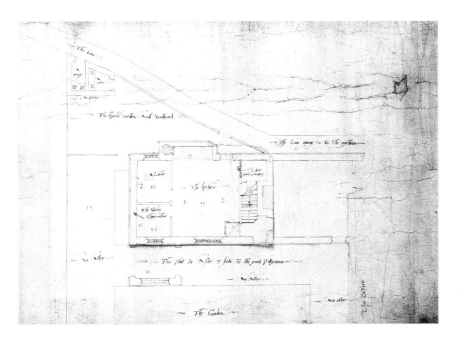

71 Cursitors Hall, London, c.1575
One of the earliest scale building plans other than those drawn for the Crown, showing alterations to be made at Cursitors Hall in Holborn. A second plan shows the first floor and an underlay the basement. A note below the centre gives the scale, '7 fots To the ynch', and the name of the architect, John Symonds.
Public Record Office, MPA 71

outside the context of royal works. Then, however, we have scale-plans of the custom house at Hull drawn by William Browne in 1573 and of works at Cursitors Hall in London by John Symonds about 1575 – drawn for officials of royal revenues and of royal justice, but not for the queen's own ministers – and, quite outside government service, a plan of the Royal Exchange in London with a scale of feet along one edge, which cannot be earlier than 1566 and may be some years later. It seems reasonable to see these as part of the widening appreciation of scale-maps that we have seen in Saxton's county maps and the earliest estate maps. Edward Worsop, in his treatise of 1582 – *Discoverie of Sundrie Errours* – refers to the scales of miles on the geographer's map, of furlongs on the surveyor's and of feet on the carpenter's plan. On the other hand, what may be the earliest building contract with annexed plans that actually survive, for a house at Woolavington in Sussex in 1586, is drawn in traditional style, to shape but not to scale and with measurements written in. The full chronology of development, besides the part that plans played in the techniques of construction, has still to be worked out.

CHAPTER SEVEN

Maps and the Law

Maps and the Law

OF the successive themes explored in this book it is the use of maps in legal contexts that we know least about. When were maps first accepted by courts as a way of setting out the layout of lands or boundaries that were in question? When did a court first order a map to be made? Did all courts come to an awareness of maps at the same time, or were courts of equity, for instance, readier than common-law courts to make use of them? All these questions are unanswered – but they are not unanswerable, and current and future research may well discover a great deal to enlarge and correct the provisional picture given in this chapter.

Law courts on the Continent were already making use of maps by the late fifteenth century. Records of the Great Council at Mechelen – principal court of the Low Countries under Burgundian rule – include sketch-maps from 1487 onwards, and by then there had been a long association of maps with law courts in France. There, Jehan Boutillier wrote in 1395 that a case put before the Parlement at Paris must be in writing, but that plans and pictures ('*figure et portraict*') could be used to make the matter clearer. Some fifteenth-century local maps from Burgundy are called *tibériades* – the word derives from the River Tiber and refers to the work of Bartolo da Sassoferrato, an Italian writer on law who in a treatise of 1355 argued the value of plans in lawsuits over rights and boundaries in rivers.

None of this can be paralleled in fifteenth-century England. Certainly some of the few local maps can be connected with particular disputes and lawsuits. The map of Inclesmoor in Yorkshire, of which there are two versions in the archives of the Duchy of Lancaster, probably originated about 1407 in litigation between the Duchy and St Mary's Abbey, York, over rights of cutting peat. The map of pastures and meadows near Chertsey Abbey in Surrey is drawn in a cartulary of the abbey beside an account of a dispute with tenants over pasture rights there. But there is nothing to suggest that any version of these maps, or any other fifteenth-century English map, was ever produced in a court of law. The dispute that the Chertsey map illustrates was in the 1420s but the map was drawn in the cartulary not before the middle of the century. Its purpose there was to make clear to succeeding generations what the dispute was about – what the abbey's claims were – and this seems to have been the motive for drawing other maps that can be connected with particular disputes. They were for reference, a safeguard for the future, clarifying and supporting the written documents so as to avoid the risk of losing rights with the lapse of time through misunderstanding or inadequate records. That lawyers too might use maps for personal reference is suggested by an early-sixteenth-century map of a mill and adjacent lands on the upper River Derwent in Yorkshire; sketched, perhaps by Thomas Nicholas of Bridlington, on a blank page in a fourteenth-century collection of statutes, it probably relates to a case about water rights. The earliest surviving English document that formally associates a map with legal process is one of 1499 that sets out the title of John Atwyll, alderman of Exeter, to a piece of land there – title that was disputed by the prior of Launceston. At the top is a plan of the property at the junction of two streets, with pictures of walls and buildings, and it may be significant that the document was written by a public notary, following Continental procedures, not working in the tradition of the English common law.

We have seen that a building contract of 1586 may well be the earliest to survive with a plan of the proposed building attached to it – but we have seen too how from 1380 onwards contracts with masons and carpenters stipulate that a building must conform to an agreed drawing or plan. Such a plan must have been considered, in effect, as much a binding part of the contract as any written clause, though we do not know whether this was ever tested in a court of law. In view of this it may seem surprising that plans were not drawn to accompany deeds of conveyance, supporting and elucidating the often lengthy written descriptions of properties that were being leased or sold. In fact two deeds of 1534, now in the archives of the Drapers' Company, may well have pioneered the use of plans in this way. By them Thomas Cromwell, the king's secretary, bought two small plots of land in London, and attached to each by the sealing tag is an outline plan of the plot, drawn to shape but not to scale, noting its dimensions and identifying neighbouring properties. On the first deed the plan

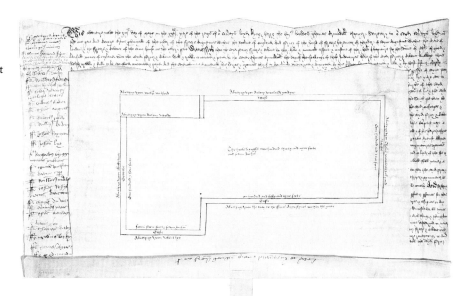

72 Plot of land in London, 1534
A conveyance from the Austin Friars to Thomas Cromwell of a plot of land adjoining the friary in Broad Street is perhaps the earliest surviving deed of this kind to set out the property on an attached plan. It is not drawn to scale, but identifies adjacent properties and gives the dimensions – 'The hoole leynght two hundred thirty and nyne foote and sevyn Inchis'.
Drapers' Company, London, deed AV 183

73 OPPOSITE Byfield and Chipping Warden, Northamptonshire, c.1550
The note bottom right shows that a copy of this plan was taken to the assizes at Northampton. The case arose when a local landowner enclosed part of the highway from Banbury (bottom right) to Byfield (top centre) and other roads. The fields in the relevant area at the centre of the map are all named and are drawn on a larger scale than the rest.
British Library, Additional MS. 63748

Maps and the Law

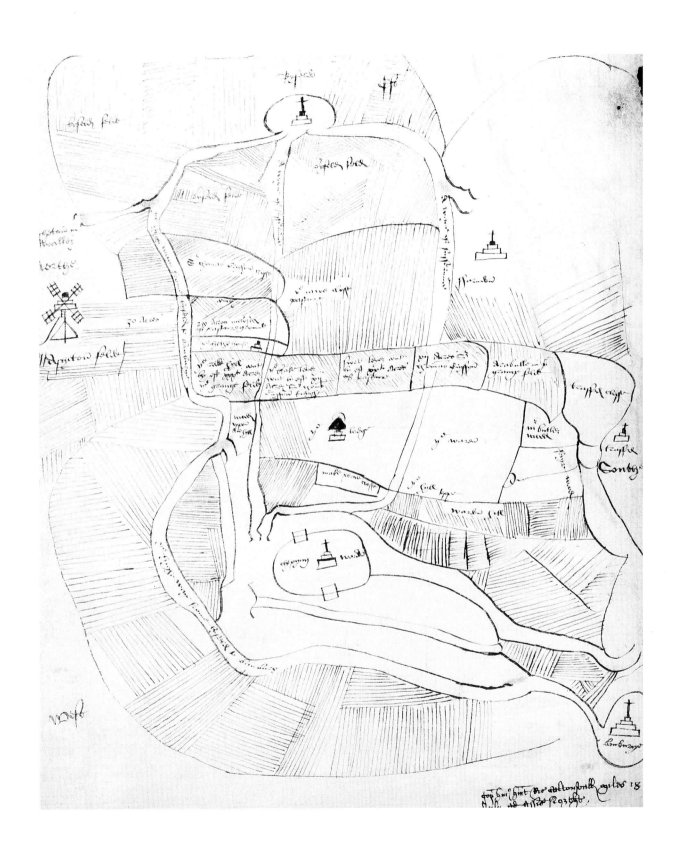

74, 75 Formby, Lancashire, mid-sixteenth century

Two relics of a case over pools in marshland west of Formby – a map and a statement of each party's claim. Both are meant for display. The map is over 3 feet square and the small letters of the statement are about ½ inch high.

74 The map has south-west at the top, though it is west according to the directions given by the head in the middle of each side. Formby is top right, Altcar on the left, and the marshland is vividly shown. All wording is the same way up – the map is to be viewed either vertically or from one side only, not round a table.
Public Record Office, MR 2

75 The statement sets out 'The situacion and distaunce of the meres as they by allegged', first by the plaintiff, then by the defendant. Each point is introduced by 'Item' in red.
Public Record Office, MPC 90

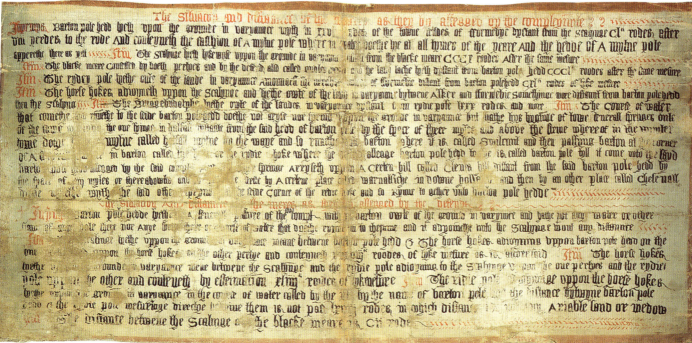

106

simply sets out graphically what is written in the document, where there is a full description of bounds and dimensions, concluding 'as ytt ys more apparatly (*i.e. apparently, visibly*) dilated in a platte indented herunto annexed'. The second deed, drawn up seven weeks later, is more venturous. The document names the parish and ward, but the property is further located only as 'lyeng & beyng in length & bredyth as in a platt indented to these presentes annexed apperith more at large' – the plan provides the only identification and description. Earlier examples of plans in conveyances may well come to light, but they are certainly extremely unusual until long after 1534.

Some fifteenth-century maps were drawn in connection with disputes about boundaries and rights over land. We have seen that from the early sixteenth century landowners might draw maps not for general reference like the estate maps late in the century but for particular *ad hoc* reasons – and disputes of this kind must lie behind many of these maps, drawn to make a point now obscure to us, to set out a claim now forgotten. We may guess that this was why the map of the mills at Wimborne St Giles in Dorset was drawn early in the century, or the map of the marshes at Cliffe in Kent at some time after 1537. Almost certainly, to take a much later example, this was the reason for a pictorial map of Gunton in Suffolk and its surrounds, extending as far as Lowestoft and its fish houses – it was drawn for one or other party to a suit in Chancery in 1578–9 over the ownership of Gunton Denes, which appear close to the centre of the map. But though in the fifteenth century we find no suggestion that maps were ever brought into a court of law, this certainly happened from early in the sixteenth century. When, in 1508, 27 persons sealed the declaration, with map attached, about the ownership of a meadow at Elford in Staffordshire, this can only have been so that it could be put before a court, and before long this was being done without the formality of a sealed document. In 1531 a case over rights of common on lands of Lord Darcy at Rothwell in Yorkshire, recently discussed by R.W. Hoyle, was being heard in a court of the Duchy of Lancaster; the court appointed commissioners to make investigations on the spot, and to make matters clear to the court they had a sketch-map made of the area in question, giving a copy of it to Lord Darcy. But the map might not be made especially for the court. A pen-and-ink map of an area in Northamptonshire, north-east of Banbury, seems to relate to a dispute over the enclosure of lands near Chipping Warden about 1550, and in one corner a note has been added 'Copia huius habuit Ricardus Saltonstall Miles 18 July ad assisas Northt' – Sir Richard Saltonstall had a copy of this at the assizes at Northampton on 18 July. This implies – but does not prove – that a landowner concerned in the dispute made the map for his own purposes but sent a copy to be shown to the court when it seemed that this would be helpful.

Unless the copy sent was a careful adaptation of the map that we have it

Maps and the Law

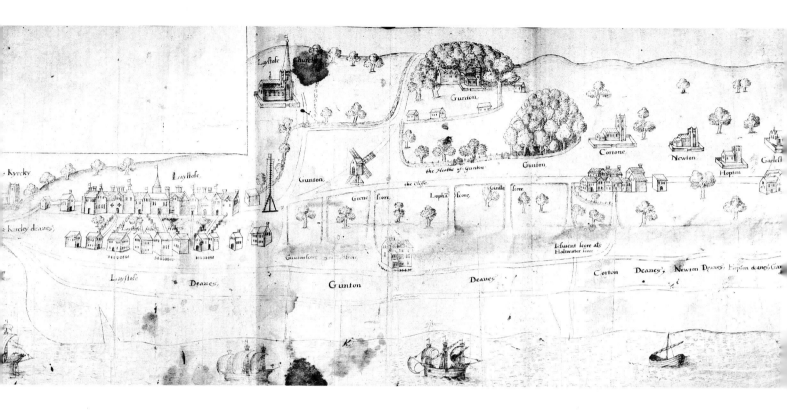

76 Lowestoft and Gunton, Suffolk, 1579–80

Drawn probably in connection with a Chancery suit over the ownership of Gunton Denes – flat land at the foot of the cliffs, reached by paths called scores – lying just below the centre of the map. On the right, the places north of Gunston and Corton – Newton, Hopton, Gorleston – are shown disproportionately close to each other. Between Lowestoft ('Laystofe') and Gunton the beacon is prominent.

British Library, Additional MS. 56070

cannot have been very suitable for general display and will have been meant for close inspection by the court. However, some maps seem to have been drawn with a view to being exhibited and seen from a distance. In 1537, in a case over land in Walland Marsh in Kent and Sussex, a deponent giving written evidence submitted a copy of a map which had been made in connection with an earlier dispute and had come into his possession some twelve years earlier; the copy is preserved with his deposition and is a quite extraordinary map. In front, at the foot of the map, are a stony road crossing a bridge with a gate beside it, all drawn large, as are also a bridge and a road at the top of the map, in the background; on the intervening grassland or marsh are two tiny pictures of buildings representing the villages of Iden and Brenzett. We may reasonably assume that what really mattered on the map – if not in 1537 then in the earlier dispute – were the roads, the gate, the two bridges and the stream flowing between them, and that these were shown large enough to be visible to all when the map was held up. fig. 81 This is only a guess – but we are on surer ground with two items, dating probably from the 1550s, which relate to Formby in Lancashire. One is a map, some 3½ feet square, painted boldly and vividly on canvas, with wording in clear black or red letters on yellow scrolls and with all detail shown large. That it was fig. 74 meant for display in court is confirmed by the other item. This is not a map at all but written statements of boundaries as claimed first by the plaintiff in a case, then by the defendant – but they are written for all to see, in painted red and black letters half an inch high on a grey and white painted background. Together they have every appearance of being relics of a visual presentation made at some point in connection with a case before a court of the Duchy of Lancaster four and a half centuries ago.

fig. 75

As the statement sets out both claims, not just one side of the case, the Formby map was probably commissioned by the court itself. This was nothing new: already in 1528 the Duchy chamber ordered 'a cart or platt' to be made of common lands in dispute at Over Haddon in Derbyshire. A picture-map drawn in 1564 for a case of disputed bounds between the manors of Allerton and Wavertree in Lancashire similarly sets out the alternative versions of the boundary, but written in panels at the foot of the map; it was to be viewed, along with other evidence, by commissioners appointed by the court who were to define the boundary. A map of Leeds and its surrounds that has been dated 1560 seems likely from its endorsement to have been commissioned by the Duchy of Lancaster chamber – the endorsement's address is notable for its polite confusion: 'To our soveringe Ladye the quene in hir honourable Courte of hir grace his courte of hir graces Duchye Chambers at Westminster'. Sometimes the court would ask local witnesses who were making written depositions to certify a map's accuracy – an example at Dedham in Essex in 1574 is interesting because the plaintiff was Thomas Seckford and one of the two deponents who answered the question about the map was Robert Mawe, who had himself been surveying Seckford's lands there.

fig. 79

There might well be cause for doubt. Three successive maps were made of Long Sleddale in Westmorland in 1578 and 1582: one ordered to be made by the Court of the Exchequer, 'a more perfecte plott' submitted by commissioners in the area, and yet another ordered by the court. A revealing case, heard before the chancellor of the Duchy of Lancaster in 1581, concerns lands in Haslingden in Lancashire. The court sent a map to seven men in the area telling them to check it and either correct or replace it. Two of them replied that they had found it so inaccurate, in misplacing or omitting some of the disputed lands, that they and one of their fellows had set about making a new one, but 'after wee hadd spente Foure or Fyve houres therein, and hadd not onlye measured the moste parte of the said landes but also sette down a platte thereof' the third man refused to continue unless certain lands, outside the terms of reference, were measured as well.

fig. 77

This was presumably the map still preserved with other papers of the case. It does not look at all like the product of a measured survey – it is roughly drawn and painted – but there is a scale of perches ('Roods') and from the map-makers' letter we may suppose that just the pieces of land under consideration

Maps and the Law

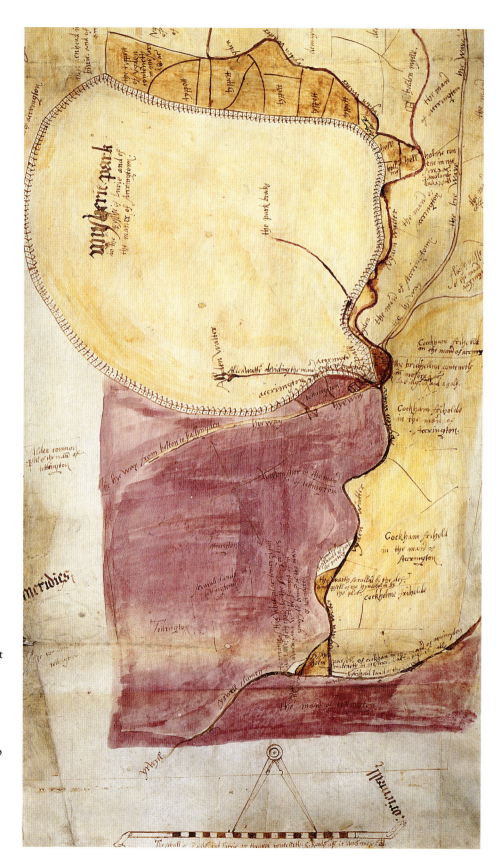

77 Haslingden, Lancashire, 1581
At first sight only a sketch-map, the scale-bar at the bottom and the attached documents show that five pieces of land in dispute – about 50 acres in all – were drawn from measured survey, the rest sketched in to set them in their local context. This is one of the earliest maps produced for a law court to be drawn to scale from measured survey. North-west is at the top and the River Irwell is at the bottom of the map; into it flows 'Ugden Watter' from the foot of Musbury Heights ('musberie park').
Public Record Office, MPC 245

Maps and the Law

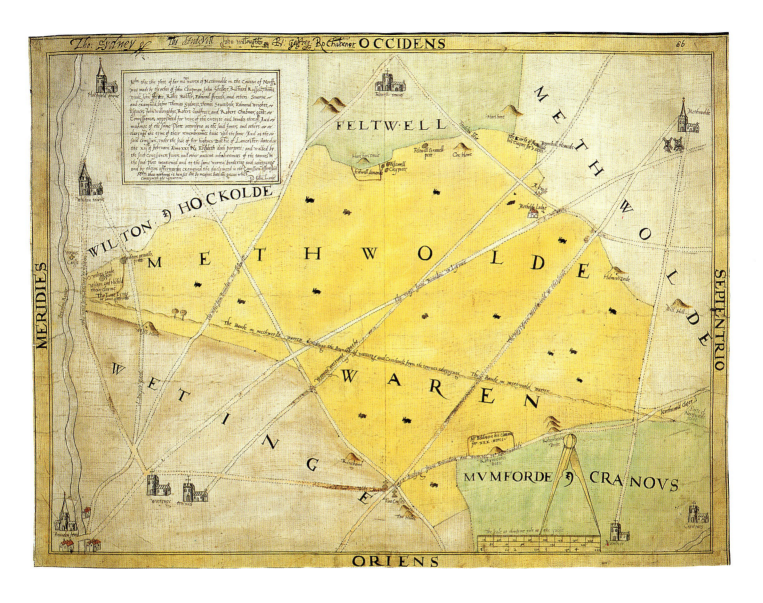

were measured and drawn to scale – some 50 acres of the 900 acres that the map covers – the rest being sketched in to place these lands in context. Two other maps made for Lancashire cases in 1580 and 1581, concerning Barnoldswick and Penwortham, are very similar in style, and the three are among the earliest surviving scale-maps that were drawn for law courts. The map from the Dedham case of 1574 is carefully drawn and may be based on a measured survey – but there is no scale-bar or other specific note of scale, and scale-maps seem to have reached the courts very suddenly about 1580. Maps from two cases over properties in Methwold in Norfolk illustrate the change. The first, of about 1566, concerned four closes, sketched roughly on a map to show their relative posit-

78 Methwold, Norfolk, 1580

Drawn by John Lane for six commissioners from the Duchy of Lancaster – five of them have signed their names in the top margin. It is noted (top left) that 'nothinge is here set out by measure but the greene which conteyneth the warren' – and it is only to this area, in fact yellow on the map, that the scale-bar applies. West is at the top; the warren, with its silhouettes of rabbits, is about 4 miles long from north to south.

Public Record Office, MPC 75

79 Leeds, Yorkshire, 1560

East is at the top and the River Aire is on the right; a bridge crosses it beside Leeds itself. Neighbouring manors are marked as belonging to the manor of Leeds. Bottom right, the black and white gable end is 'The auncyent Manor House of leedes callid Castyll Hill'. Drawn for a case in the Duchy of Lancaster chamber, the map is typical of those produced for law courts before scale-maps by surveyors were introduced in the 1580s.

Public Record Office, MPC 109

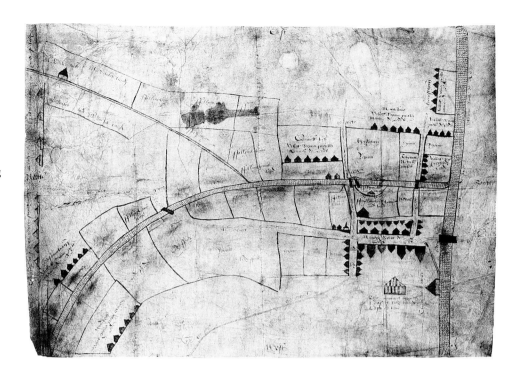

80 Old Byland, Yorkshire, by Christopher Saxton, 1598

An area some 4 by 2 miles, mapped for the Court of the Exchequer to show boundaries and common pasture disputed by two neighbouring landlords. The court ordered the map to be made 'at the equall chardge of both the said parties'. In the title, top right, Saxton identifies the colours: red for enclosures, yellow for common lands, green for a boundary. By the end of the century a map for a law court was normally like this, drawn to scale by a trained surveyor from measurement on the ground.

Public Record Office, MPB 32

OVERLEAF

81 Walland Marsh, Kent and Sussex, early sixteenth century

Like many picture-maps, drawn to no consistent scale, this emphasises features considered important by enlarging them. It was certainly drawn for a lawsuit, but what was at issue – what was the significance of the gate, the two roads and the two bridges – has not been discovered. The sea – to the south – is at the top of the map. Between it and the road in the foreground lies the whole of Walland Marsh, with tiny pictures marking the villages of Brenzett and Iden.

Public Record Office, MPA 61

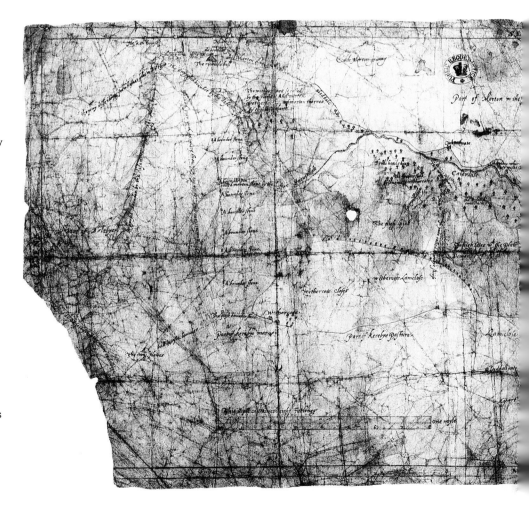

ions; the second in 1580 was over a warren, and this appears on a map with silhouettes of rabbits, but also with an elaborate scale-bar and a warning note that 'nothinge is here set out by measure but the greene which conteyneth the warren'.

fig. 78

It seems, then, that while surveyors were persuading landowners to let them draw maps of their estates, they were also persuading the courts that the best maps to use were measured maps drawn by surveyors. In both cases the popularity and widespread appreciation of Saxton's county maps probably contributed to their success. But though in the 1580s the law courts were simply following a general trend in their map-making they may have taken the lead in earlier stages of development. There is still much to be discovered about the maps drawn for lawsuits in the first half of the sixteenth century, and it is here that we may well find the earliest and most significant innovations of Tudor cartography.

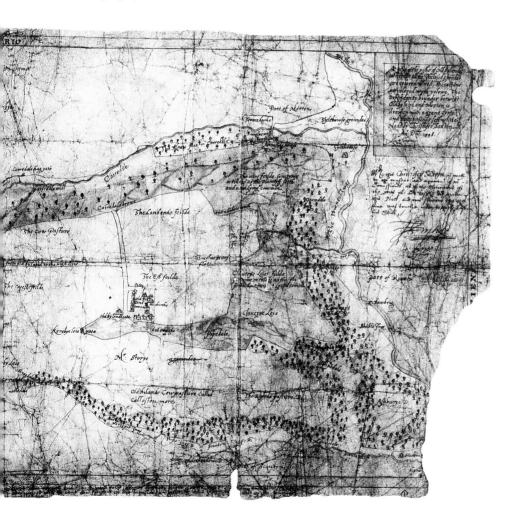

Further Reading

The only general book on the precise field of cartographic history examined here is Sarah Tyacke and John Huddy, *Christopher Saxton and Tudor Map-making* (London, British Library, 1980), which is brief but wide-ranging and well illustrated. However, the essays in *English Map-making 1500–1650*, ed. Sarah Tyacke (London, British Library, 1983), cover many of its aspects in more detail, and particularly relevant are the introduction by Sarah Tyacke, 'Meaning and Ambiguity in Tudor Cartography' by J.B. Harley, 'Italian Military Engineers in Britain in the 1540s' by Marcus Merriman, 'Three Elizabethan Estate Surveyors' by Peter Eden and 'Christopher Saxton's Surveying: an Enigma' by William Ravenhill. The medieval precursors of these Tudor maps are described in P.D.A. Harvey, *Medieval Maps* (London, British Library, 1991), chapter 6, and, in greater detail which includes an account of medieval surveying, in *Local Maps and Plans from Medieval England*, ed. R.A. Skelton and P.D.A. Harvey (Oxford, Clarendon Press, 1986). R.A. Skelton and J. Summerson, *A Description of Maps and Architectural Drawings in the Collection made by William Cecil, First Baron Burghley now at Hatfield House* (Roxburghe Club, 1971), is a detailed and fully illustrated description of the largest collection of sixteenth-century English maps outside the British Library and the Public Record Office, and it includes many references to maps elsewhere. Catherine Delano-Smith and Elizabeth Morley Ingram, *Maps in Bibles 1500–1600* (Geneva, Libraire Droz, 1991), provide a full list and discussion of this important group of maps. Arthur M. Hind, *Engraving in England in the Sixteenth and Seventeenth Centuries. Part I: The Tudor Period* (Cambridge, Cambridge University Press, 1952) is valuable for reference on all engraved maps. Victor Morgan, 'The Cartographic Image of "The Country" in Early Modern England', *Transactions of the Royal Historical Society*, 5th series, 29 (1979), pp.129–54, is an important account of the growing perception of cartography.

For fortifications and the maps associated with them, *The History of the King's Works*, ed. H.M. Colvin, vols 3 and 4 (London, H.M. Stationery Office, 1975–82), is essential. L.R. Shelby, *John Rogers: Tudor Military Engineer* (Oxford, Clarendon Press, 1967), is an important study of a key figure. P.D.A. Harvey, 'The Portsmouth Map of 1545 and the Introduction of Scale Maps into England', in *Hampshire Studies*, ed. John Webb, Nigel Yates and Sarah Peacock (Portsmouth,

Further Reading

Portsmouth City Records Office, 1981), pp.32–49, gives a fuller account of these topics than is possible here. There are important descriptive bibliographies of the maps from Plymouth and Portsmouth: Elisabeth Stuart, *Lost Landscapes of Plymouth: Maps, Charts and Plans to 1800* (Stroud, Alan Sutton, 1991), and D. Hodson, *Maps of Portsmouth before 1801* (Portsmouth Record Series, vol.4, 1978). A.H.W. Robinson, *Marine Cartography in Great Britain* (Leicester, Leicester University Press, 1962), is the standard work on marine charts.

On the use of maps in defensive schemes and more generally by the government, P. Barber, 'England I: Pageantry, Defense and Government: Maps at Court to 1550' and 'England II: Monarchs, Ministers and Maps 1550–1625', in *Monarchs, Ministers and Maps*, ed. David Buisseret (Chicago, University of Chicago Press, 1992), pp.26–98, is crucially important and touches on many other aspects of the use of maps in this period. David Marcombe, 'Saxton's Apprenticeship: John Rudd, a Yorkshire Cartographer', *Yorkshire Archaeological Journal*, 50 (1978), pp.171–5, first brought to light Rudd's role in Tudor cartography, and on Saxton himself Ifor M. Evans and Heather Lawrence, *Christopher Saxton: Elizabethan Map-maker* (Wakefield and London, Wakefield Historical Publications and Holland Press, 1979), is a comprehensive and definitive work. Saxton's county maps are reproduced in *Christopher Saxton's 16th Century Maps*, ed. W. Ravenhill (Chatsworth Library, 1992), and Speed's in *The Counties of Britain. A Tudor Atlas by John Speed* (London, Pavilion Books in association with the British Library, 1988). E.M.J. Campbell, 'The Beginnings of the Characteristic Sheet to English Maps', *Geographical Journal*, 128 (1962), pp.411–15, introduces the development of conventional signs and of keys to them. J.H. Andrews, 'Geography and Government in Elizabethan Ireland', in *Irish Geographical Studies*, ed. Nicholas Stephens and Robin E. Glasscock (Belfast, Department of Geography, The Queen's University, 1970), pp.178–91, provides a brief general account of Irish mapping in this period.

The fullest account of the 1553–9 map of London is Martin Holmes, 'An Unrecorded Map of London', *Archaeologia*, 100 (1966), pp.105-28. R.A. Skelton, 'Tudor Town Plans in John Speed's *Theatre*', *Archaeological Journal*, 108 (1952), pp.109–20, includes a discussion of Speed's precursors.

E.G.R. Taylor, 'The Surveyor', *Economic History Review*, 1st series, 17 (1947), pp.121–33, is a first exploration of the surveyor's role down to 1600. Maurice Beresford, *History on the Ground* (London, Lutterworth Press, 1957), demonstrates the value of, especially, estate maps as evidence of the Tudor landscape. P.D.A. Harvey, 'Estate Surveyors and the Spread of the Scale-Map in England, 1550–80', *Landscape History* 15 (1993), gives an account, fuller than here, of the emergence of the estate map.

There is no published work specifically on either plans of buildings or the use of maps in law courts in the sixteenth century

Index

Adams, Robert, 45, 52, 65, fig.33, fig.40
Africa, 7
Agas, Ralph, 21, 74–5, 80–1, 83, 85, 89, 91, 93, 95, fig.60–61
Aire, River, fig.6, fig.79
All Souls College, Oxford, 93
Allerton, Lancashire, 109
Ambleteuse, Picardy, 31, 37, 52
Ames, Israel, 80, 85, fig.58–9
Andwell, Hampshire, 21, 79, 95, fig.8
antiquaries, 25, 61, 64, fig.12–13
Arcana, Archangelo, 29
Ardres, Picardy, 29
Armada, 37, 41, 45, 65, fig.33, fig.40
Armas, Duarte de, 28
artists, 28, 37, 43, 74
atlases, 52, 54, 58, 84
Atwyll, John, 104
Austria, 98

Banbury, Oxfordshire, 107, fig.73
Barber, Peter, 44
Barnoldswick, Lancashire, 111
Bartlett, Richard, 65
Bath, Somerset, 77
beacons, 58, fig.13, fig.22, fig.26–7, fig.76
Beeston, Cheshire, 77
Belchamp St Paul, Essex, 81, 85, fig.58
Benese, Richard, 80
Bergamo, Antonio, 28
Berwick-upon-Tweed, 32, 37
Bible, 7–8, 24
Blackwall, Middlesex, 45, fig.33
Bologna, 28
Borough, William, 41
Boulogne, 28–9, 31, 37
Bourne, William, 83
Boutillier, Jehan, 103
Bow Bridge, Middlesex, 47
Bowes Castle, Yorkshire, 47
Brancepeth, Co. Durham, 47
Braun, Georg, 74–5
Brenzett, Kent, 108, fig.81
Bridlington, Yorkshire, 103
Brighton, Sussex, 43–4, fig.27
Bristol, 67, 75, 77
Britain, maps of, 8–9, 52–4, fig.2
Broughton Craig, Angus, 29
Browne, John, 65
Browne, William, 101
building plans, 7, 95–101, 104, fig.66–7, fig.70–71
Bullock, Henry, 45, fig.29
Burghley, Lord, *see* Cecil, William
Burgundy, 103
Byfield, Northamptonshire, fig.73
Byland, Old, Yorkshire, fig.80

Caius, John, 75
Calais, 31, 97, fig.21, fig.66
Calstock, Cornwall, 79, fig.55
Cambridge, 16, 75, 77, fig.7
Canterbury, 75, 77; St Augustine's Abbey, 25
Carlisle, 29, 31, 74, fig.50

Carrickfergus, Co. Antrim, 95, fig.68
Catholic recusants, 48
Cecil, William, Lord Burghley, 47, 48, 52, 54, 65, 76–7, fig.34, fig.36, fig.38, fig.48
Chancery, Court of, 107, fig.76
Charles V, Emperor, 44
charts, nautical, 41, fig.23; *see also* portolan charts
Cheney, Lord, 41, 89, 95, fig.22, fig.60–61
Chertsey, Surrey, fig.5; Abbey, 21, 103, fig.5
Cheshire, 77
Chester, 75, 77
Chipping Warden, Northamptonshire, fig.73
chronicle, local, 67
city plans, *see* town maps
Cliffe, Kent, 79, 107, fig.54
Cologne, 74
Colvin, H.M., 27
compass, 13
conventional signs, 13, 61
Cornwall, 60; coast, 44–5, 67, fig.32
Corpus Christi College, Oxford, 93
Cotehele House, Cornwall, 79, fig.55
county maps, of England, 9, 52, 54, 58, 60–61, 64–5, 84, 91, 101, 113, fig.36–9
Coventry and Lichfield, bishop of, 27
Cromwell, Thomas, 44, 97, 104, fig.72
Cumberland, fig.29
Cuningham, William, 13, 73

Da Sassoferrato, Bartolo, 103
Da Treviso, Girolamo, 28–9
Da Vinci, Leonardo, 28
Darcy, Lord, 107
Dart, River, fig.32
Debatable Lands, 45, 47, fig.29, fig.34
Dedham, Essex, 83, 109, 111, fig.57
deeds of conveyance, 104, 107, fig.72
Deptford, Surrey, 95
Derwent, River, Yorkshire, 103
Devon, coast, 44–5, 67, fig.32
Dewsbury, Yorkshire, 54
Digges, Leonard, 37, 83
Digges, Thomas, 33, 37, 83, fig.15
Ditchley, Oxfordshire, frontispiece
Don, River, fig.6
Dorset, coast, 44, fig.26
Dover, Kent, 28, 31–3, 37, 41, 83, fig.15, fig.17–19
Drapers' Company, 104
Drayton, Michael, 7
Dugdale, William, 25
Duisburg, 9, 53
Dunningley, Yorkshire, 54
Dürer, Albrecht, 28, 31
Durham, 53–4, fig.1

Edinburgh, fig.4
elevations, 28, 98, fig.65
Elford, Staffordshire, 27, 79, 107, fig.14
Elizabeth I, Queen, 7, 24, 47, 53–4, 58, frontispiece; arms of, 74, fig.7, fig.37–8, fig.42
Elmham, Thomas, 25
Elyot, Thomas, 43
enclosure, 89, fig.63, fig.73
engineers, military, 12, 24, 28–9, 31–3, 37, 41, 43, 64, 98, fig.15, fig.69
England, maps of, 53, 91, fig.12; *see also* Britain; county maps
English Channel, 65, fig.40
Enniskillen Castle, Co. Fermanagh, fig.43
Esk, River, fig.29
Essex, 48
estate maps, 24, 65, 79–93, 101, fig.58–64
Eton College, Buckinghamshire, 95
Exchequer, Court of, 109, fig.80
Exeter, 44, 75, 104

Falmouth, Cornwall, fig.30

Faversham, Kent, 43–4
Ferrers, Sir John, 27
Ferrybridge, Yorkshire, fig.6
Fitzherbert, John, 80
Formby, Lancashire, 108–9, fig.74–5
fortifications, 24, 27–41, 45, 52, 67, 98, fig.20, fig.25, fig.44, fig.47, fig.49–50
France, 12, 41, 43–4, 103; king of, 44, 97, fig.66
Frisius, Gemma, 13

Garrard, Sir William, fig.64
Gateshead, Co. Durham, fig.47
Genebelli, Federico, 37
Germany, 7, 12, 43, 98
Gheeraerts, Marcus, frontispiece
Greenwich, Kent, 44
Gregory, Arthur, 37
Guines, Picardy, 29, 31, 37, fig.20
Gunton, Suffolk, 107, fig.76
Gussage All Saints, Dorset, 27, fig.16
Gussage St Michael, Dorset, 27, fig.16

Hackney, Middlesex, 47
Haddon, Over, Derbyshire, 109
Halton, Cheshire, 77
Hamond, John, 75
Hampshire, fig.39
Harwich, Essex, 31, 44
Haschenperg, Stefan von, 29
Haslingden, Lancashire, 109, fig.77
Hastings, Sussex, 28
Hatton, Sir Christopher, 17, 89, fig.62–3
Haulbowline Fort, Co. Cork, fig.44
Hawthorne, Henry, 97, fig.67
Henry VIII, King, 17, 28–9, 43–4, 98
heralds, 25, 61
Hertford Castle, 95
Hertfordshire, 61
Higham Ferrers, Northamptonshire, 77
Hogenberg, Frans, 74–5
Hogenberg, Remigius, 75
Holbein, Hans, 44
Holdenby, Northamptonshire, 89
Hooker, John, 75
Horman, William, 95
Hoyle, R.W., 107
Hull, 31, 37, 44, 67–8, 98, 101, fig.24–5, fig.69–70
humanists, 43
Humber, River, 41, fig.6, fig.12

Iden, Sussex, 108, fig.81
Inclesmoor, Yorkshire, 21, 103, fig.6
Ireland, 12, 52, 65, 77, fig.2, fig.41–3, fig.68
Irwell, River, fig.77
Italy, 25, 27–8, 31, 43, 67; books published in, 7, 30, 65; Italian artists, 28; Italian military engineers, 28–9, 37, 98
itinerary-maps, 8

Johnson, Rowland, 37

Kelso, Roxburghshire, 29
Kent, 58, 91, 93, fig.13, fig.37; coast, 43–4, fig.28, fig.54
key, to conventional signs, 13, 61, fig.39
Kingswear, Devon, fig.32
Kirby, Northamptonshire, 89, fig.62–3

Laleham, Middlesex, fig.5
Lambarde, William, fig.13
Lambeth, Surrey, 95
Lancaster, Duchy of, 23, 103, 107, 109, fig.6, fig.10–11, fig.78; chamber, 109, fig.79; chancellor, 109
Lands End, Cornwall, 44, fig.32, endpapers
Lane, John, fig.78
Launceston, Cornwall, prior of, 104
law courts, maps in, 12, 24, 103–13
Lea, River, 47

Index

Lee, Richard, 31-2, 98, fig.20
Lee, Sir Henry, frontispiece
Leeds, Yorkshire, 109, fig.79
Leigh, Valentine, 83
Lexham, West, Norfolk, 80, 85
Leysdown-on-Sea, Kent, 43
Liddel Water, Liddesdale, 47, fig.29, fig.34
Lily, George, 9
Lizard, Cornwall, endpapers
London, 21, 22, 24, 45, 47, 64, 74-5, 95, 104, fig.33, fig.51, fig.53; Broad Street, fig.72; Cursitors Hall, 101, fig.71; Lombard Street, 97; London Bridge, 21, 95, fig.53; Royal Exchange, 101; St Paul's Cathedral, 81; Tower of London, 21, fig.53
Lough Neagh, Cos Armagh and Tyrone, fig.42
Louvain, 13, 43
Low Countries, *see* Netherlands
Lowestoft, Suffolk, 107, fig.76
Lyne, Richard, 17, 75, fig.7
Lythe, Robert, 65

Margate, Kent, 43-4
Mawe, Robert, 83, 109
Mechelen, Great Council at, 103
medals, 7
Mediterranean, 7-8
Melcombe Regis, Dorset, fig.26
Mercator, Gerard, 9, 53-4
Merioneth, 58
Methwold, Norfolk, 111, 113, fig.78
Middlesex, 58, 61, fig.37
mills, 27, 61, 85, 103, 107, fig.3, fig.5, fig.16, fig.22, fig.41, fig.55
Mogeely, Co. Cork, 65
Monmouthshire, 23, 58
Montgomery, 58
Montreuil, Picardy, 28
Moravia, 29
More, Thomas, 43
Mountjoy, Lord, fig.42
Mount's Bay, Cornwall, fig.31, endpapers
Munster, fig.41
Musbury Heights, Lancashire, fig.77

Naples, kingdom of, 25
Netherlands, Low Countries, 12, 31, 37, 41, 61, 103
New College, Oxford, 93
Newark-on-Trent, Nottinghamshire, fig.3
Newcastle-upon-Tyne, 29, 76, fig.47
Nicholas, Thomas, 103
Norden, John, 60-61, 77, 91, 93, fig.39
Northampton, 107, fig.73
Northamptonshire, 77, 107
Northumberland, 48, fig.36
Norwich, 68-9, 74-5, 77, fig.46, fig.52
Nowell, Lawrence, 25, 52, 54, fig.12, fig.35
Nuremberg, 28

Orléans, Charles, duc d', 21
ornament on maps, 24, 74, 84
Ouse, River, Yorkshire, 69, fig.6
Owen, George, 65, 84
Oxford, 75, 93

Paignton, Devon, fig.32
Paris, Parlement at, 103
Parker, Matthew, fig.7
Pembrokeshire, 65
Pendennis Castle, Cornwall, fig.30
Penwortham, Lancashire, 111
Penzance, Cornwall, fig.31
Peterborough, Northamptonshire, 77
Picardy, 12, 32, 37, 44
plane-table, 13, 84, 93
playing-cards, 7, 64-5
Plumberow, Essex, fig.59
Plymouth, Devon, 37, 45

Popinjay, Richard, 37
Portland, Dorset, fig.26, fig.40
portolan charts, 7-9
Portsmouth, Hampshire, 37, 41, 45, 69, 73, fig.49
Portugal, 28-9
Poulter, Richard, fig.23
Poyntz, Fernando, 37
printed maps, 8, 12, 21, 58, 65, 74-5, fig.7, fig.36-8, fig.40, fig.51, fig.53; printing plates, 74; proofs, 52, 54, 76
Privy Council, 54
Ptolemy, 7, 9, fig.2

quadrant, 83
Queenborough Castle, Kent, fig.22

Raleigh, Sir Walter, 65
Ravenhill, William, 58
Ricart, Robert, 67
roads, routes, 21, 31, 47, 60-61, fig.8, fig.73
Rogers, John, 17, 31, 33, 37, 98, fig.20, fig.69
Rome, 9
Rossetti, Giovanni di, 29
Rothwell, Yorkshire, 107
routes, *see* roads
Rudd, John, 52-4, 93
Rye, Sussex, 28

Salisbury, bishop of, fig.56
Saltonstall, Sir Richard, 107
sanctuary, places of, 68-9, fig.46
Saxton, Christopher, 8-9, 17, 24, 52-4, 58, 60-61, 64-5, 76-7, 84, 91, 93, 101, 113, fig.13, fig.36-8, fig.64, fig.80
Scala, Gian Tomasso, 29, fig.47
scale of maps, 37, 58, 65, 84; drawing maps to scale, 8, 12, 15-16, 27, 29, 31-3, 37, 41, 45, 64-5, 69, 73-5, 83-4, 91, 98, 101, 111, fig.20-1, l fig.25, fig.29, fig.49, fig.70-71, fig.77
Scarborough, Yorkshire, 44, fig.45
Scilly Isles, 52
Scotland, Scottish border, 9, 12, 45, 47, 54, fig.2, fig.4, fig.29, fig.34-5
Seasalter, Kent, fig.28
Seckford, Thomas, 17, 24, 54, 93, 109, fig.37
settlement, new, 31, 65, fig.21, fig.41
Severn, River, fig.48
Seymour, Edward, earl of Hertford, fig.4
Shakespeare, William, 7
Shelby, L.R., 33
Shell Ness, Kent, 43
Sheppey, Isle of, Kent, 32, 41, 43, fig.22
Sherborne, Dorset, fig.56
Shrewsbury, Shropshire, 76, fig.48
Sicily, 52
Sittingbourne, Kent, fig.64
Sleddale, Long, Westmorland, 109
Smith, William, 25, 61, 77, fig.52
Somerset, coast, 44
Southwark, Surrey, 8, 69, 74
Spain, 28
Speed, John, 61, 77
St Ives Bay, Cornwall, endpapers
St Mary, Scilly Isles, 52
St Mawes Castle, Cornwall, fig.30
St Michael's Mount, Cornwall, fig.31, endpapers
Suffolk, 54
Surrey, 58, fig.37
survey, court of, 93
surveyors, 12-13, 31, 60, 80-81, 83-5, 89, 91, 93, 113; surveyors' sketch-maps, 83, fig.57
surveys, written, 54, 79-80, 83, 85, 89, 91, 93
Sussex, 58, fig.37
Swaleness, Kent, 41
Symonds, John, 101, fig.71

Tamar, River, fig.55
Tame, River, fig.14
tapestries, 7, 64

Tartaglia, Niccolo, 31
Thames, River, 45, 79, fig.33, fig.53-4; estuary, 21, fig.23
Thanet, Kent, 25, 32, 37, 43
theodolite, 13, 83-4, 93
Thérouanne, 44
Thomas, John, fig.43
Thompson, John, 32, 37, fig.18
Tilbury, Essex, 45
Tirrell, Edmund, 81, 85, fig.59
Toddington, Bedfordshire, 21, 89, 93, 95, fig.60-61
Toledo, treaty of, 44
Tor Bay, Devon, fig.32
Totnes, Devon, fig.32
Tournai, 44
town maps, 21-2, 25, 64, 67-77, fig.45-53
treatises, legal, 103; on fortification, 28, 31; on surveying, 80, 83-4, 91, 93, 101
Trent, River, fig.3, fig.6
Treswell, Ralph, 17, 89, fig.62-3
triangulation, 13, 58, fig.13
Tynemouth, Northumberland, 29, 76
Tyrone, earl of, fig.42

Ubaldini, Petruccio, 65
Ulster, 65

Venice, 31
Volpe, Vincenzo, 28, 32, fig.17

Wales, 9, 12, 52, 54, 58, 64-5, 84, 91, fig.2
Walland Marsh, Kent and Sussex, 108, fig.81
Warden, Chipping, Northamptonshire, 107
Wards and Liveries, Court of, 54
Wark-on-Tweed Castle, Northumberland, 29
Wavertree, Lancashire, 109
Wentworth, Richard, 31
Westminster, 74, 109
Weymouth, Dorset, fig.26
White Castle, Monmouthshire, 23, fig.10-11
Whitstable, Kent, fig.28
Wight, Isle of, fig.39-40
Wimborne St Giles, Dorset, 27, 107, fig.16
Winchester, Cathedral, 98, fig.65; College, 95
Windsor, Berkshire, 47; Castle, 97, fig.67
Woodbridge, Suffolk, 54
Woolavington, Sussex, 101
Woolwich, Kent, fig.33
world maps, 7
Worsop, Edward, 83-4, 101

Yarmouth, Great, Norfolk, 21-2, 76, fig.9
York, 68-9; St Mary's Abbey, 103, fig.6
Yorkshire, 52, 91, 93, fig.12; coast, 41